The Next Tsunami

The Next Tsunami
Living on a Restless Coast

Bonnie Henderson

OREGON STATE UNIVERSITY PRESS • CORVALLIS

The paper in this book meets the guidelines for permanence and durability of the Committee on Production Guidelines for Book Longevity of the Council on Library Resources and the minimum requirements of the American National Standard for Permanence of Paper for Printed Library Materials Z39.48-1984.

Library of Congress Cataloging-in-Publication Data
Henderson, Bonnie, author.
    The next tsunami : living on a restless coast / Bonnie Henderson.
        pages cm
    Includes bibliographical references and index.
    ISBN 978-0-87071-732-1 (alk. paper) -- ISBN 978-0-87071-733-8 (e-book)
    1. Tsunamis--Pacific Coast (America) 2. Tsunami hazard zones--Pacific Coast (America) I. Title.
    GC220.4.P37.H46 2014
    363.34'940979509146--dc23
                                2013040039

Oregon State University Press
121 The Valley Library
Corvallis OR 97331-4501
541-737-3166 • fax 541-737-3170
www.osupress.oregonstate.edu

This is a work of nonfiction. Conversations not witnessed by the author have been reconstructed with the collaboration and approval of those participants still living.

Seismologists use a variety of scales to characterize the relative size of an earthquake, including the local magnitude or Richter scale, developed in the 1930s, and the surface wave magnitude scale. The United States Geological Society now uses the moment magnitude scale, developed in the 1970s, to estimate all modern large earthquakes. Unless noted, mention of earthquake magnitude in this work is based on the moment magnitude scale.

Table of Contents

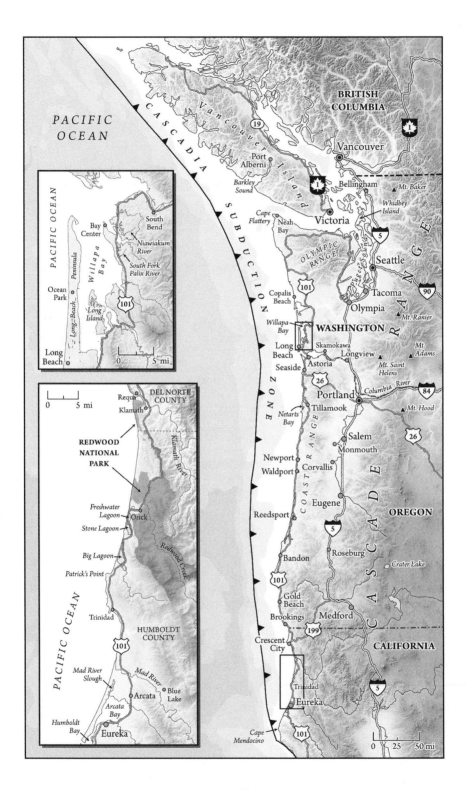

Introduction

I FIRST MET TOM HORNING IN 2008 on a visit with my brother in Gearhart, Oregon. We had stopped for coffee and muffins at Pacific Way Bakery. In one corner a tall man stood, reclining against the wall, long arms clasped over his chest, holding forth good-naturedly on a range of topics, from the floodlights that shone on the beach and inhibited viewing of the night sky (an apparent pet peeve) to the pros and cons of a proposal in neighboring Cannon Beach to rebuild city hall as a tsunami evacuation building.

Randall waved Tom over to our table and we introduced ourselves. By the time we left the bakery, my head was swimming with the story I'd heard: of a man who, as a ten-year-old boy in Seaside, Oregon, had narrowly escaped being swept to sea when a tsunami rushed into the mouth of the river where he lived, a man who now parlayed his bona fides as a professional geologist and hometown boy to stir up interest in preparing for the much larger tsunami threatening the town, all the while continuing to live at what amounted to, in Seaside, Tsunami Ground Zero: the mouth of the river.

I began poking around, educating myself in the history of plate tectonics theory and the 1964 tsunami from Alaska. I was astonished at the pace of change that had occurred in the earth sciences in just the past half-century. Satellites were being launched into space well before humans had a clear understanding of what processes actually formed the valleys and mountains, the rifts, ranges, and deep ocean trenches of our own home planet, what caused it to periodically tremble and convulse, its seas to catastrophically overflow. I was also struck by the synchronicity of the science and by the almost uncanny correspondence between the course of Tom's life and the development of what we know of earthquake and tsunami risk on the Pacific Northwest coast. Not that Tom did the research himself, but he seemed to know everyone who did. (Just weeks before we met, Tom later told me with chagrin, he had accidentally broken the bow seat in Brian Atwater's aluminum canoe during a field trip with the esteemed geologist.)

Seaside, Tom's hometown, is just one of many communities from Vancouver Island to northern California threatened by a major tsunami from an earthquake at a fault line scientists call the Cascadia Subduction Zone. But it is arguably the town most at risk, due to the size of its population and the lay of the land. Shortly before I met Tom, the U.S. Geological Survey published a paper rating the tsunami vulnerability of towns on the Oregon coast. In chart after chart, the bar for Seaside stretched the longest, with neighboring Cannon Beach and Gearhart generally close behind and sometimes slightly ahead. Most acreage and percentage of developed land in the tsunami inundation zone? Seaside. Highest number of people living in the inundation zone? Seaside. Greatest number of schools, and child care centers, and clinics, and hotels, and churches in the tsunami zone? Seaside, Seaside, Seaside. In fact, Seaside's vulnerability to every category of tsunami mayhem is so high it makes the city, statistically speaking, an outlier. Similar USGS reports for Washington and California followed. The Washington coast, like the west coast of Vancouver Island, is much less developed than the Oregon coast; except for Aberdeen, Washington's most vulnerable towns are very small, and Aberdeen has accessible high ground to run to. That state's most vulnerable shoreline state parks are on the Strait of Juan de Fuca, far east of the fault line, giving vacationers an hour to evacuate rather than, as on the outer coast, fifteen or twenty or thirty minutes. Crescent City, California, is about the same size as Seaside, but most of its residents live on high ground, as do most citizens of Eureka and smaller towns on Humboldt Bay.

"How can you write a book about a disaster that hasn't happened yet?" a friend had asked me. Fair question, and one I couldn't answer at the outset. But the world continues to turn, and the tectonic plates under us to shift and grow, about as fast as your fingernails grow: slowly, in human terms, but not imperceptibly so. Ultimately Acts of God (and, apparently, of small-town arsonists) provided the book's denouement.

From the start of my research, I knew this story wasn't really about a future disaster. It was about the quest to understand the geologic history of the Earth, and of one corner of the Earth in particular, a place Tom Horning refers to with love and irony as Tsunamitopia. And it is about how we choose to deal with that knowledge, and that uncertainty, as a society and as human beings with competing priorities and limited imaginations. Humans like Tom Horning, going to sleep and waking up in a beautiful place steeped in childhood memories and knitted together with lifelong relationships. A place he couldn't think of leaving.

Prologue
Litany
MARCH 10, 2011

IT WAS PAST 10 P.M. WHEN THE CHIME SOUNDED on the computer speakers in Tom Horning's office, which he hoped to lock up and leave well before midnight. The next meeting of the board of the North Coast Land Conservancy was scheduled for the following afternoon, and as secretary, Tom was expected to show up with minutes from the previous month's meeting. An e-mail at 10 p.m.? Probably spam, or churnings from a fellow night owl. Nothing urgent, certainly, but one can't help but wonder. Curiosity overcame him.

He relaxed. The sender was "USGS ENS," the Earthquake Notification System of the United States Geological Survey. The subject: "(M 7.9) NEAR EAST COAST OF HONSHU, JAPAN." So not spam, but nothing urgent. Still, they were always interesting, these alerts that arrived almost daily, sometimes several in one day: constant reminders of the disquietude of the Earth. Tom had set his notification parameters so he would be alerted only by earthquakes in the Pacific Ocean region, minimum magnitude 7.5. Across the earth there are more than one hundred detectable earthquakes every day, some days more than two hundred, one place or another, most of them too small to be felt by humans. Tom knew it generally takes an underwater earthquake of at least magnitude 7.5 or so to generate a tsunami. And tsunami-generating earthquakes were the only kind Tom was really interested in.

Magnitude 7.9? It might be enough to trigger a tsunami, but not much of one. He clicked to open the message and scanned the stats. "Depth: 24 km"—*shallow focus, right at the leading edge of the subduction zone.* "Local standard time in your area: 21:46:23"—*just a half hour ago. Mid-afternoon in Japan, March 11.* "Location with respect to nearby cities: 81 miles E of Sendai, Honshu, Japan." *Good thing it was just magnitude 7.9, or those poor people would be underwater by now.*

Tom had anticipated a late night when he arrived at his office after 8:30 that evening. Such are the dues of procrastination and passion and, perhaps, over-commitment in the affairs of a small town. Tom's approach to preparing the meeting minutes was characteristically thorough, painstaking even; they were practically a transcript of the meeting, captured on a digital recorder, leaving nothing to chance. He leaned back in his heavy oak office chair, springs squeaking, to peer out his west-facing office window.

The headquarters of Horning Geosciences consisted of a single cluttered room on the ground floor of a small, two-story office building—just right for a one-man operation—off U.S. Highway 101 at the north end of Seaside, a dozen miles south of the Columbia River on the northern Oregon coast. Across the highway, beyond a slice of ebony water where Neawanna Creek emptied into the wide Necanicum River estuary, he could see a light on in the kitchen of his own small white house, gray in the moonless night and hunched just above the waterline at the end of the peninsula south of the estuary. No light escaped the living room windows, but the wooden blinds were probably closed. This time of night his wife, Kirsten, and father-in-law, Mike, were almost certainly in that room, watching television.

Some evenings Kirsten worked at her printmaking, gouging grooves in blocks of wood, chipping away the negative space between what would become, once the block was inked and pressed onto paper, a shore pine's needles or the vanes on a kingfisher's wing. But at the end of a day spent working at the community college in Astoria, television in the living room was her after-dinner default. TV is what usually occupied Mike, except when Kirsten or Tom got him out of the house for a drive or a meal at a restaurant. He didn't remember anything of what he saw on TV, but he seemed to enjoy the parade of images.

Lights from other living rooms and kitchens shone like sparks in the otherwise dark neighborhood, but no moonlight penetrated the clouds that night to sheen the estuary or frost the ocean breakers beyond. Poking the bridge of his glasses with a finger, Tom swiveled back to the monitor. *Well, no time like the present,* Tom thought—maybe even said out loud to himself, as he sometimes did, alone in the office. Two hours, he figured. Three, max.

Tom had moved his office out of the back bedroom of his house and into this small business complex just a few years earlier. It was where he tended to do his volunteer work as well as his *work* work—geologic hazard assessments, mostly, for homeowners and prospective homeowners and business owners and sometimes governmental bodies seeking to avoid landslides, slumping, or other unpleasant surprises. Everything he needed was here, and here everything fit him: the office chair, the large oak desk with its legs raised on blocks, the bookshelves stretching almost to the ceiling, the side tables within comfortable reach of his long arms. At six feet, four inches, Tom was a tall man, solidly built. All was arranged for his convenience, in heaps mostly, strategically placed: the parks and recreation district's business in one clump (he was a member of that board too), city documents in a couple of other piles (he chaired the city's planning commission), his books and journals and maps, his computer-monitor-printer-fax assemblage, buttes of unshelved reports, a drift of unfiled memos at their angle of repose. The door and walls were plastered with posters advertising recent nature walks with the land conservancy, some led by Tom, and brightly colored geological maps, one nearly as tall as Tom himself. Those maps revealed at a glance—if you knew what you were looking at—the story of uplift and subsidence and tension and compression across North America from the collision of tectonic plates and related forces of nature across millions of years. A very old story, one as clear to Tom as the nightly news; nearly as predictable, eminently more sensible. Working at night, alone in his office, Tom not only had access to all the resources he might conceivably need, but he was subject to a minimum of distractions. So when the phone rang ten minutes later, he was momentarily startled.

"Tom, it's Al"—Al Smiles, director of the Seaside Chamber of Commerce, his voice strained with urgency. The introduction was unnecessary. No one else with that crisp Welch accent would be calling Tom at that time of night, or maybe ever. "There's just been a big earthquake in Japan, and they're expecting a tsunami. I'm watching CNN. It sounds pretty serious."

"Oh, yeah, I just saw that," Tom replied laconically. People worry so much, he thought. They don't realize how many earthquakes go off every day, all over the globe. Devastating for the people on the ground, of course. All a part of life on Earth.

"Are you watching the television?"

"I'm in my office, Al," Tom replied, patiently, almost teacherly. "No TV here. A tsunami, huh? Well, I don't expect it will be much of one. The quake wasn't particularly big."

"Well, we'll see. Let's hope not."

Odd, Tom thought, hanging up and turning back to the computer to refocus on the meeting minutes. Al Smiles had spent twenty-three years in the British army, fifteen of them in special ops. He had been the personal security advisor for a dot-com billionaire before following his fiancée to Seaside, Oregon. Now, when he wasn't figuring how to lure more tourists and retirees and businesses to Seaside, he volunteered with Clatsop County Search and Rescue, rappelling down sea cliffs to reach injured surfers or clueless hikers stranded by the incoming tide. *Takes a lot to rattle Smiles.*

Not ten minutes later the phone rang again.

"Tom, you'd never guess." It was Smiles again, his voice even more urgent. "I'm watching it on CNN. They were saying, 'We're expecting a tsumami,' and then they cut to this scene of a port full of boats, and then this wave arrived and the boats were all getting pushed under this big bridge, and there were cars driving along the waterline getting picked up ..."

"So they've had a little local wave action there in Japan?" Tom asked, humoring Al, meanwhile checking the computer for updates. Nothing— just a preliminary magnitude 7.9 earthquake off Japan. It was a good spot for a bad earthquake, certainly. Right on the Japan Trench, where the northwestern edge of the Pacific tectonic plate collides with an arm of the North American Plate—the Okhotsk Microplate, some geologists called it. *Fifty years of plate tectonics,* he mused, *and we still know less about how our own Earth works than about the locations of dwarf galaxies 200,000 light years away.* But Tom had faith in the numbers—and more confidence in himself than in news reports from the other side of the globe. Whatever Al thought he was seeing, it couldn't be that bad, not from a magnitude 7.9 quake.

"Oh my God, Tom, you're not going to believe this," Al continued, "it's just absolute devastation, they're showing this wave going up this big flat-bottomed valley and it's taking everything out with it, houses and cars, pushing this debris—my God, there's a building on fire floating on top of the water, I've never seen anything like it, it's just complete devastation ..."

"Oh, really?" Not that Al was prone to exaggeration, but magnitude 7.9? That was hardly big enough to produce even a small tsunami. "I'll have to watch it later at home."

"I wish you could see this, Tom. You wouldn't believe it, oh my God ..."

"Well, I'll be up for a while, Al. Call anytime."

Tom checked the clock in the corner of the monitor: past 11 p.m. He'd barely made a dent in the land trust minutes. *Might not get home by midnight after all.*

Not until about 11:30 did the phone ring again.

"Hey, Tom, you should see this footage they've been showing on the news." It was Kirsten. "It's pretty startling."

"Yeah, well, it can't be too bad. It wasn't much of an earthquake, all things considered."

"It looks pretty big on TV. The video shows some pretty big shaking, stuff falling off buildings, and now this tsunami pouring over seawalls and picking up cars and houses ..."

"I'll check it out. I've just got to finish these minutes, then I'll be home. Might be after midnight, at this point."

"What do you think? Should we be worried here?"

"I can't see why. I haven't seen anything yet to indicate we'll get a tsunami here. But I'll keep an eye on it."

"OK. I may stay up."

"Don't wait up for me."

She stifled a laugh. "I won't."

Tom turned back to the computer, but clicked away from the land trust minutes and over to the Internet, where he started opening familiar websites: the USGS, the National Oceanic and Atmospheric Administration, and its West Coast and Alaska Tsunami Warning Center. There it was, unchanged: preliminary magnitude 7.9, accompanied by a "Tsunami Information Statement" for California, Oregon, Washington, British Columbia, and Alaska—*information* being NOAA-speak for the lowest level of alert, really no alert at all. "NO tsunami warning, watch, or advisory is in effect for these areas," it read. *Pretty emphatic* no, he thought. *Good thing, that emphasis. People do tend to overreact.*

A *distant* tsunami—that's the possibility NOAA was referring to, and the term coined for it. The tsunami Japan was apparently experiencing, if there really was a tsunami, would be a *local* tsunami, kicked up by an earthquake off Japan's own shore. The greatest damage from a tsunami like that would be to the nearby shore: to Japan itself. But tsunamis are also long-distance travelers. They cross entire oceans at the speed of a jet plane, losing power incrementally but—depending on the severity of the quake and other factors—maintaining the capacity to do damage thousands of miles from where they started. At Seaside, Oregon, for instance, and the rest of the west coast of North America. Especially Seaside, and places like it: Crescent City, California, and Vancouver Island's Clayoquot Sound and Alberni Inlet— flat coastal plains, and bays and fjords where an incoming tsunami, distant or local, gets squeezed and its effects get magnified. If this quake in Japan did stir up a tsunami big enough to cross the Pacific Ocean to Oregon, it wouldn't be the first time. Seaside's fire chief and police chief and city manager were probably huddling right now, Tom figured, assessing the danger and figuring whether or when to activate the city's new, improved tsunami warning sirens scattered around town.

But magnitude 7.9? It was possible geologists had initially underestimated the magnitude of this thing, he mused. He opened a screen listing the coordinates of the fault zone off Sendai, Japan. If the quake *were* big enough to generate a significant tsunami, it would already have begun crossing more than 4,000 miles of Pacific Ocean, headed east-southeast toward California and Mexico, he calculated from the angle of the trench. Not aimed directly at Seaside. But Seaside might catch the edge.

Then a new e-mail popped up, this one from an acquaintance of Tom's, a land use planner in Astoria. The e-mail added nothing to what Tom already knew and seemed, to Tom, to be jumping the gun a bit: "FYI everyone. Big earthquake and tsunamis in Japan right now. No official watch or warning for our area yet, but I am certainly texting my Warrenton, Gearhart, etc., friends."

Tom turned back to the land trust notes. He managed to focus for another ten minutes or so before until another chime sounded, from the same acquaintance. The subject line: "FW: TSUNAMI WATCH ISSUED."

Tom knew the language. *Watch* was NOAA's official term for "danger level not yet known; stay alert," and was just one step above *Information* ("minor waves at most; no action suggested"). *Watch* was followed, in order of urgency, by *Advisory* ("strong currents likely; stay away from the shore"). At the top of the hierarchy was *Warning*: "inundating waves possible; full evacuation suggested."

The land trust minutes could wait. Tom clicked away from that screen to the NOAA website—they were still calling it a magnitude 7.9 quake, which was genuinely odd—then opened the CNN website streaming live footage from Japan. Al was right; this was no minor tsunami. Then shortly after 12:30 a.m., he checked NOAA's West Coast and Alaska Tsunami Warning Center site again.

Bingo. There it was, the update he by now expected: magnitude 8.9, an entirely different ball game. Not just one-eighth bigger than a magnitude 7.9, but—according to the logarithmic scale used to quantify earthquakes' energy and potential for mayhem—thirty-two times bigger. An *enormous* quake, one of the biggest on record anywhere in the world, if this new magnitude estimate held, just shy of the 2004 Sumatra-Andaman earthquake, which spawned a tsunami that killed more than 230,000 people and wreaked havoc along coastlines throughout southeast Asia. This quake off northern Japan was, by any standards, a Big One. Big enough to kill a lot of people in Japan. Big enough, probably, to send some kind of measurable wave to North America. Possibly even to Seaside.

It would be an interesting night, and morning, and that thought made Tom smile. There would almost certainly be TV reporters calling him from Portland, and live interviews with reporters who would be, even now, scrambling to gather a crew and make the hour-plus drive to the coast to film the action and get commentary from Tom Horning, Local Geologist and Tsunami Expert. If he could unearth his old camcorder from under the piles of papers and miscellany in his office, he'd be able to videotape the wave himself, documenting its entrance into the Necanicum River from his favorite tsunami-watching spot: the rise at the end of Twenty-Sixth Avenue, two blocks from his house and directly east of the bay mouth. Distant tsunamis don't hit Seaside every day, not even once a decade, on average, at least according to official reports. These distant tsunamis were always exciting: another opportunity to add data points to his on-going, lifelong study of tsunami behavior in Seaside, Oregon.

Exciting, but not alarming. This was not the tsunami Tom spent at least part of every day thinking about, ranting to the city manager or planning director about, e-mailing politicians and fellow geologists and members of Seaside's Tsunami Advisory Group about. A distant tsunami like this was a double-edged sword, he figured: it got people in Seaside thinking about tsunamis, and that was a good thing. But it got people thinking the wrong way, thinking that they're no big deal, that if you buy new warning sirens, you're done, you've finished the job.

The tsunami Tom did worry about would make warning sirens irrelevant. Minutes before that tsunami hits, the sirens will probably be silenced by the force of the earthquake, which he fully expected will cut the electricity even as it manufactures a tsunami the likes of which no one had witnessed in North America for hundreds of years.

Three hundred and eleven years, to be exact. And two months.

He hoped he wouldn't miss it, frankly—felt, in a way, that he was entitled to witness it, given the nearly twenty years he had spent learning about it and preparing for it and trying to engage others in Seaside to prepare for it. He was fairly confident he and Kirsten would survive it, suspected that his father-in-law and many of his neighbors would not. He already knew what he would say to the TV reporters when they started calling, any moment now: "... a distant tsunami, not the same as a local tsunami, really not dangerous as long as you stay off the beaches and out of harbors ..." Knew what he wished he could say, to them and to every city official who kept pushing tsunami readiness to the bottom of the to-do list.

The land is talking to you all the time! he would tell them. *It's an interesting language. If God talks to us, he does it through natural forces. And this litany of stuff that we're living through, the earthquake in Chile, and now in Japan? It's telling us that we're going to have one here pretty soon, and you better be ready. Really, what the hell is wrong with you that you can't hear these messages?*

Tom swiveled the heavy chair to take another look out the window, at the point of light that, despite the reflections cluttering the window's surface, shone like a tiny beacon across the highway, beyond the black estuary: the light in his own kitchen window.

I've never had God really talk to me except through geology, Tom mused—maybe even spoke the words, to no one but himself.

It's loud and clear. I wish other people would listen.

1

The First Tsunami

Tom, 1964

It was called Good Friday, but to a ten-year-old boy living at the mouth of the Necanicum River on Oregon's north coast, every Friday was good. It took Tom Horning fewer than ten minutes to pedal home from Central School in downtown Seaside to the Venice Park neighborhood where he lived, across the bay from the outlet of Neacoxie Creek. He would have at least three hours to play outside until dinnertime. After dinner there would be TV in the living room with the family, maybe even *The Twilight Zone* if his mom was in an expansive mood, and then the whole weekend lay ahead. This day—March 27, 1964—could actually be characterized as a great Friday. For one thing, the weather was perfect: sunny, shirtsleeve warm, not much wind. Not at all typical for late March in Seaside, where early spring was often drizzly at best or, at worst, lashed with sou'westers that pummeled the shoreline with huge waves and littered yards with the downed limbs of Sitka spruce and shore pine. But not this weekend. The forecast called for more fair weather, at least through Saturday, which meant a dry Easter egg hunt at the city park.

It was a fairly straight shot north up Holladay Drive, along the east bank of the river, past the old wooden Fourth Avenue Bridge (condemned to cars but a handy shortcut for kids on bikes) and the bridge at Twelfth Avenue, past the high school and its playing fields, then across Twenty-Fourth Avenue, where Tom veered onto Oregon Street. The houses here were modest, the yards neat with blooming daffodils and spike-leafed crocosmia or strewn with Japanese glass floats and other beachcombing treasures, the vacant lots bristling with a tangle of salal and blackberry and spruce. Two blocks farther, at Twenty-Sixth Avenue, he turned right and pedaled the last block to his house. He ditched his bike in the garage, picked up his new Tonka dump truck, and headed back outside.

The truck was perfect: bright yellow cab with real windows, a yellow hinged dump box just like a real dump truck, black wheels and grill and bumpers, and not a scratch on it. Tom had had it only nine days, since his tenth birthday. He grabbed the truck and started walking west, up Twenty-Sixth Avenue toward the broken lines of ocean breakers he could see over the top of the bluff, two blocks away. He was headed toward his favorite spot in the world: the edge of the bay where the Necanicum River mingled with the waters of Neawanna and Neacoxie creeks before sliding into the Pacific Ocean. At the top of the bank he resumed excavation operations, using sticks and his hands to dig in the hard, cemented sand layer just below the top of the bank, where the brushy shoreline vegetation ended and the sandy riverbank began.

Tom sat cross-legged, leaning into the bank and squinting in concentration, his mouth set slightly open with a sense of expectancy. He was a stocky, gap-toothed boy already as tall as many adults, his brown hair cut short and ragged by his mother's scissors. Most of his friends already had Tonka trucks, many since the first grade. He had yearned to have what the other boys had. That desire had gnawed at him, had led him finally to beg his mother for it. And this year she'd given in. Maybe the cancer had overridden her usual frugality—cancer and weeks of absence in Seattle, recovering from surgery. Bobbie Horning wasn't the most demonstrative of moms, not big on hugs. She had other ways of showing her affection. A truck, for instance.

Now that he had it, he didn't know quite what to do with it. Another boy might have put the yellow dump truck to work transporting sand from one locale to another. Tom had at his disposal an unlimited inventory of soft river sand just steps away, a whole bay full of it. Beyond lay Oregon's widest ocean beach, composed of sand transported over millennia to the ocean via the broad Columbia River, which meets the Pacific some seventeen miles to the north. But the firm upper riverbank was what drew him: a place not so much to work the truck as to park it. The compact sand here had just the right structural integrity for single-truck garages, hollowed-out sand caves whose walls, he wagered, wouldn't easily collapse in wind or rain.

The Hornings' house was one of a small collection of houses that made up the north Seaside neighborhood of Venice Park—aptly named, with water on three sides. Tom's house was the last one on Pine Street, at the very end of the peninsula formed where Neawanna Creek, running north, curved west and met the broad estuary at the mouth of the Necanicum River, also running north. Ocean, river, creek: water in one form or another defined Seaside. The wide ocean beach was what drew the tourists and turned the town crazy in summer, easily doubling its population of 4,000. Spring break was even wilder, keeping the police busy corralling partying college students. But tourism was what had put Seaside on the map almost a century earlier, with its holiday hotels and tent camps by the seashore, and tourists of all stripes were what kept it going in 1964 as well. The motels and restaurants and chowder cafés and bars. The souvenir shops and salt water taffy. The arcades and bumper cars and Ferris wheel. The aquarium with its octopus and jellyfish and harbor seals, performing for the tourists and applauding their own antics with slick gray pectoral fins.

To get to the beach from the highway running through town, you first had to cross one of a half-dozen bridges over the Necanicum River, which bisected downtown. The Necanicum tumbles west out of the Coast Range, calming to a slow meander when it reaches the coastal plain, veering north when it hits the base of 1,150-foot-tall Tillamook Head, the mountain of basalt defining the south end of town, then meeting the creeks and swinging west a scant hundred yards to join the sea. You hardly noticed the river running through town, for all the restaurants and shops crowding its banks, until you needed to go east or west through downtown and had to find a bridge. Neawanna Creek flows parallel to the Necanicum just a few blocks to the east, more or less defining the city's eastern edge. It serves to separate the town from the rolling patchwork of growing trees and clear-cut forest that dominate the view to the east, forest nurtured by the rain that drizzles, or showers, or drips, or dumps much of the winter and spring.

Tom and his four siblings—two older sisters, an older and a younger brother—had moved with their mother to the house at the edge of the bay five years earlier, after their father had died. Tom had been born at Seaside Hospital and spent his first five years at Crown Camp, a logging camp and company town in the forest three miles east of Seaside. John Horning— educated in architecture but lacking a degree—worked as a purchasing agent and sometimes as facilities designer for Crown Zellerbach, the biggest player in the leading industry in Clatsop County, which was cutting down trees and turning them into lumber and paper products. Crown Camp consisted of about twenty houses and offices clustered on three lanes set deep in the woods where the Coast Range runs down to meet the Clatsop Plains. Tom and his three older siblings had had the run of the place. Identical white frame houses with little yards and laundry lines and sandboxes. A jungle gym and a baseball diamond. Gravel roads with logging trucks grumbling incessantly by. Surrounding Crown Camp were deep woods where massive trees grew out of the hulking stumps of other, older trees, and the ground was a sponge of moss overlain with knobby tree roots, and red-legged frogs hid at the edges of muddy seeps shaded by salal and huckleberry. Everyone knew everyone else at Crown Camp. There were drunken fathers and sober fathers. Suspendered fathers with muddy hardhats carrying chainsaws with grimy hands, and fathers in white shirts and clean hardhats carrying blueprints. There were mothers with crossed arms and defeated faces. Mothers with open doors, and sometimes cookies. Bobbie Horning was different from most of the other mothers at Crown Camp, mothers who wouldn't allow their children to leave the yard on their own or to venture past the last house or beyond the ball field, lest they get lost in the woods or run over by a log truck. *I took the laissez-faire approach,* she would tell people with a laugh. *I figured with five kids, I could afford to lose one or two.* And everyone would laugh with her when she said it, even her own kids, because they all knew it wasn't true.

In the summer of 1959, just after Tom had turned five, his father had been transferred to Crown Zellerbach's office in downtown Portland, and Tom and his four siblings—baby David had been born the previous October—traded Crown Camp and the infinite forest for a house in the Portland suburb of Milwaukie. Just three months later, Tom's father entered the hospital with a bleeding ulcer on his esophagus and didn't come home. John Horning, whose hemophilia ruled out rough work outdoors, who couldn't risk cutting his finger, started to bleed inside. Doctors transfused him again and again, a special blood drive was held, 265 pints of strangers' blood was pumped into his veins and was filtered through his kidneys before bleeding back out. It took nearly a week for him to die. Eventually his blood started to clot, but by then, his kidneys had worn out.

There wasn't much outward grieving. Bobbie had suddenly become a widow with five children ages ten and under. The kids tucked their grief and confusion inside, and Bobbie focused on moving forward. There was never any question about where they would live; all of their friends were back in Seaside, either at Crown Camp or in town. Bobbie's first task was to find a job there. She had trained as a registered nurse, though most of the nursing she'd done was with her own kids. She started asking around and found that Dr. Russell Parcher, the general practitioner who had delivered both Tom and David, was looking for an office nurse, and he hired her. Next, she needed a house for the six of them. Bobbie found one for sale on the

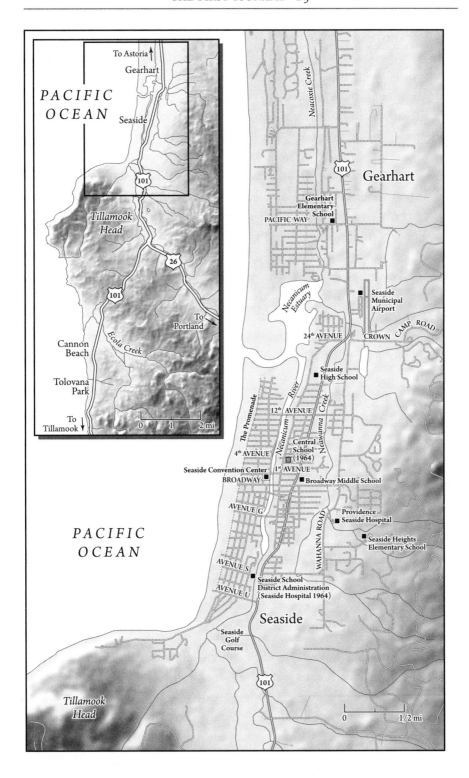

Necanicum River in the middle of town, just blocks from Dr. Parcher's office and Central School, a convenient location. Then she heard about another house several blocks north, at Twenty-Sixth and Pine. It was farther from downtown but just a short walk from the high school, which the kids would all eventually attend. And as at Crown Camp, there was plenty of room for children to run and play outdoors, minus the log trucks.

Tom would never forget the details of their return to Seaside that day in September 1959. By the time Bobbie had packed the last box and loaded baby David and five-year-old Tom into the station wagon and left Portland, it was already late afternoon. She had barely pulled into the driveway when Tom jumped out of the car and ran to catch up with his older siblings, who had arrived earlier with family friends. They had already been under the house and inside the old detached garage and across the yard to the little two-story cottage and out onto the river beach, beyond the white plank fence that scribed the boundary between the bay and the lawn surrounding the house. The tide was rising, hissing quietly as its leading edges crept over the hard white sand flats. Tom could see his sisters standing on a dock built on pilings over the river at the edge of the property, and he ran over to join them. He walked to the end of the dock and stood there, surveying his new realm. To the west was the broad estuary, swelling with the tide that carried lopsided clumps of sea foam the color of buttermilk, and beyond them the white breakers marking the river's meeting with the Pacific Ocean. To the north, past the flat expanse of tidewater and salt marsh where narrow Neacoxie Creek slipped southward into the bay, he could see lights starting to twinkle in the houses above Little Beach, in the community of Gearhart. His eyes traced the curve of Neawanna Creek eastward, upstream, where the creek narrowed, flowing lazily alongside the highway.

By now his sisters were lying face-down, peeking through cracks in the dock to watch the water below, and he joined them, flattening his face against the damp, splintery planks. The water that evening was crystal-clear, and as his eyes adjusted, he began to see schools of silvery shiner perch flashing in the water. Below the perch, milling along the bay's sandy bottom, were the dark brown bodies and fleshy whiskers of bullheads.

"Look!" Lynne whispered, nudging him. Through the clear water, they could see a jumble of old logs, discarded metal tanks, and a broken boat rudder, all submerged by the tide and colonized by clumps of white barnacles. In the fading light, Tom could make out feathery appendages emerging from the little fortress-like barnacle shells: wispy fingers, urgently raking the water with rhythmic intensity. "See! You can see them gathering in food!" she said.

Tom had never seen anything like it.

Never—not even at Crown Camp—had he been in a place quite so magical, and so alive. With every shift of the tide, every minute even, something changed, something new revealed itself. Even the bay itself was constantly changing, its channels and pools continuously reshaped by storm and tide. In the summer he would join the other neighborhood boys skinny-dipping on the ebbing tide, bathing in the sun-warmed freshwater pools until someone would shout *ocean water,* meaning the tide had turned and the cold Pacific—55 degrees at best—had started streaming in. Then they would make their way to shore and pull on their clothes and haul the rowboat off the lawn and go crabbing, combing the sandy bottom with garden rakes until a big Dungeness crab would grab the tines with muscular pincers, clinging all the way into the boat. Alone or with Pat, his best friend, Tom spent his fair-weather days outdoors, crabbing and fishing, watching the waves spilling onto the bay shore or soldiering up the river with the tide. They built driftwood forts and dug in the sand to build sand castles, sand volcanoes, sand moats, digging, designing, compacting, destroying, until his mother called him inside for dinner. And sometimes, spring through fall when the days were long, he'd be back outside even after dinner.

THIS NIGHT, HOWEVER—Good Friday 1964—he stayed in. *International Showtime* had started at 7 p.m., and the whole family was watching TV—everyone but five-year-old David, who was already in bed, and Bobbie, who was around the corner in the kitchen, doing the last of the dinner dishes. Lynne, a freshman at Seaside High, and her boyfriend, Lyle, a junior and a football player, were snuggling in an armchair; they'd just come in from a romantic float in the family's rowboat, riding the incoming tide under the full moon. Judy, Chris, and Tom filled the sofa in the compact living room, with Muffin, the family shepherd mix, curled at their feet. On the TV, tigers were jumping through flaming hoops at a circus, urged on by a tall ringmaster in tails and a black top hat. Now and then he'd crack a long, black whip on the ground, directing the big cats this way and that. The German audience was applauding rhythmically—*clap, clap, clap, clap, clap,* nothing like the chaotic applause Tom was used to at, say, football games or school assemblies. It was strange, the mechanical clapping. Tom reached down to scratch Muffin's back.

Then the circus scene vanished and a newscaster's head and shoulders appeared. "We interrupt this program to bring you a special bulletin," anchorman Chet Huntley intoned. "A massive earthquake struck the state of Alaska approximately 45 minutes ago, at 5:37 p.m. Alaska Standard Time. At least five people are known dead, and property damage is extensive. The main street in Anchorage has been completely flattened, and the city is in darkness. Tidal waves have reportedly struck several coastal towns, but few details are available, as communication with the region has been almost entirely cut off. Civil defense authorities are warning coastal communities to the south to prepare for a possible tidal wave."

The image on the screen switched from Huntley's face to a map of the wide northern Pacific Ocean. Alaska hunched in the upper right, looking a little like an elephant's head with the chain of Aleutian Islands its trunk, stretching west in a broad arc. Down the left side of the TV screen were the islands of Japan, and on the right side, the west coast of North America. Concentric curves—the presumed path of the tidal wave—were depicted fanning out across the Pacific, marching southward from Anchorage. The right tip of one curve ended in a scattering of islands off the Alaska panhandle. Another, south of it, touched cigar-shaped Vancouver Island. And one touched the Oregon coast just below the indentation indicating the mouth of the Columbia River, at a bump on the map that Tom knew well, the place where the long, straight coastline was abruptly broken by the jutting bulk of Tillamook Head.

Tidal wave? Tom looked around. None of his siblings seemed to be the least bit alarmed. *Tidal wave, here?* Well, Tom *was* a little alarmed. Someone should tell Mom at least. He got out of his chair and scrambled into the kitchen. "Mom? There's going to be a tidal wave," he told her matter-of-factly, standing next to her at the sink, looking up at her face in profile.

Bobbie didn't respond, not right away. She continued to work at the dishes, running a soapy sponge around a greasy plate rim, turning the plate under a stream of water from the faucet, lodging it in the dish drainer to drip, reaching for another plate. Finally, eyes still on the sink full of dishes, she spoke. "Why don't you go to bed," she said. "Chris too."

Her response struck him as odd—and patently unfair. It wasn't even bedtime yet, not even weekday bedtime, and here it was Friday night. Tom turned, a little puzzled, a little defeated—and a little relieved. He had done his duty. He had alerted his mother about the life-threatening disaster that Chet Huntley himself said was, at that very moment, bearing down on them. But if she wasn't concerned, why should he be? There had been many tidal wave warnings in Seaside in recent years. Sometimes they even blew the siren at the fire station. Nothing ever came of these warnings. He'd never actually seen a tidal wave. Nobody had, not in Seaside. And Tom recognized his mother's tone. There was no point in arguing with her about bedtime, early or late, tidal wave or no tidal wave.

He headed to the bathroom to brush his teeth, jostling with eleven-year-old Chris for room at the sink, then followed Chris out the back door and a dozen steps across the dark yard. Since the previous summer, both boys had slept upstairs in the little cottage between the house and the river, freeing up space in the main house, which was getting more crowded with two teenage girls in the family. Their Grandfather Baker—their mother's father—had visited from California for a few weeks the previous summer and had taken it upon himself to update the cottage and make it habitable. It had been built back in the 1930s, probably as a summer rental. There were entrances on both the river and the house side—porches that a previous owner had enclosed to shut out the often-harsh coastal weather. A little half bath occupied part of the south entrance, closest to the house, the entrance the boys generally used. Grandfather had taken out the decrepit old toilet and replaced it with a new one. He had also installed a shower, and he had hooked up toilet and shower to the city sewer system, which had only recently reached the houses in Venice Park. He had also run an underground electrical line out to the cottage so the boys could have heat in their bedroom. As for the north-side porch, the Horning kids themselves had turned it into what they called the "hamsterium," a little room where their half-dozen pet hamsters (and one rabbit) could run free. They'd shoved an old steamer trunk against the inside of the north entrance; people could step over it, but it kept the animals from escaping when the kids came in and out through that door.

Tom could hear the rustling sounds of hamsters nibbling their sunflower seeds and burrowing under the linoleum floor as he passed the hamsterium and headed up the wooden staircase in the dark, could smell the familiar smells: the perfume of cedar shavings mingling with the sour ammonia odor of rodent pee. At the top of the stairs, he could see the estuary through the north window, black water streaked with a sheen of moonlight, and the dark void of spruce forest bordering Neacoxie Creek on the other side of the bay. Now and then the twin spots of headlights appeared at the forest's edge and moved down the highway that followed the bayside. He pulled off his shoes and socks and pants and sweatshirt and crawled between the familiar cold sheets, curling into a ball, feeling the weight of the blankets and waiting for his own body heat to warm the bed.

Tom, Tom, wake up, Tom. It was a man's voice. He felt a hand on his shoulder, shaking him awake in the dark, reeling him up from the soft, muffled depths. It was Lyle, Lynne's boyfriend, his big hand rough, and urgent.

"Hey, Tom, wake up. There's been a tidal wave," he was saying. "Come on, let's go. There's been a tidal wave. Let's go to the house."

Tom bolted upright. "I knew it! I knew it!" he cried, instantly awake. Across the room, Chris was sitting up and rubbing his eyes. "Let's go, boys," Lyle reiterated, moving toward the stairs in the dark.

The three of them started down at a trot, Tom and Chris still in their T-shirts and underpants. The wooden stairs were cool on Tom's bare feet. Two steps above the landing where the staircase made a turn, Tom first felt dampness underfoot. From there to the landing and down the final three steps everything was wet. When he reached the bottom, his feet felt not the slick of the cottage's linoleum floor but the grainy burr of wet sand, a thick layer of it. Lyle led the way out over the sand-covered floor, through the open inner door, and stepped over a white board that was now inexplicably plastered across the bottom of the outside entrance at a slight angle, clinging with nails driven into the siding at either side. It looked like one of the planks from the fence at the edge of yard, above the river. Tom followed Lyle over the rail and onto the lawn.

But there was no lawn. What he saw instead was like something from a dream: familiar and eerily unfamiliar at the same time. Every inch of what had been green grass was now covered with dark, wet sand. The sand was strewn with debris. Some of it was the same stuff Tom was accustomed to seeing in the wrack line on the beach after a storm: seaweed and grasses and driftwood of all sizes. But there was more. Thick, lustrous yellow foam clumped in piles a foot or more thick all over the sand, and more of it piled deeper against the house and cottage. And fish: flounders and perch and bullheads, lying still and scattered all over the yard. And little waxy, translucent pink shrimp the length of Tom's pinky finger, thousands of them, everywhere.

Tom took a few steps, then stopped and stared, surveying the known world. The house was there, and the old two-story detached garage, but the concrete block patio wall on the north side of the house had collapsed. The fence was gone, its posts still standing but the flat white planks all missing— all but the one that had apparently fastened itself across the cottage entrance. The dock, too, had vanished. So had the rowboat that Lynne and Lyle had left upside down on the lawn after their moonlight float a few hours earlier. An outdoor rabbit hutch that had stood along the west wall of the cottage now staked a claim in the middle of the yard. The chicken coop, too, had drifted west but was still upright, the chickens apparently still alive, judging from the low rustling sounds that came from within. East of the cottage, drift logs lay scattered on the yard, huge logs, one as wide as Tom was tall. Beyond the north edge of the yard, past the shore pines, the river was running out fast and full like a winter flood tide, laden with dark, angular, moon-washed objects—more logs and other floating debris Tom couldn't quite make out. Across the short expanse of sand-covered yard between the cottage and the house, he could see his mother standing outside the back door, watching them, her arms crossed over her chest in the cool of midnight, hugging herself. His brothers and sisters were all out wandering on their new backyard beach too, smiling and laughing as if a little drunk on it all. And there was Mr. Jensen, their bachelor neighbor, in his long white nightshirt and white nightcap, his bare legs thrust into leather work boots, a quizzical expression on his face as he went about poking the toe of his boot into piles of debris. There was a tang in the air, a briny smell like low tide, fresh and pungent, mixed with the faint odor of rotten eggs.

Tom's mouth, which had been hanging open, now spread into a wide smile. He drew in a breath and felt his shoulders fall back, his arms reaching wide. "I knew it!" he said, over and over. "I knew it! I knew it!" He began to turn in place, and soon he was running in big circles, the balls of his feet slapping the wet sand, finding their own way among the pink shrimp, the slick, green seaweed, the buttery foam. Around and around he ran, laughing, his arms open and stretching to the very tips of his fingers, as if to embrace it all: the glistening sand, the shrimp, the logs, the late hour, the moonlight, whoever or whatever had granted him this night, this best night of his whole life.

LIKE MOST SEASIDERS, Tom's family had been taken entirely by surprise, unless you count the TV news bulletin that had interrupted *International Showcase* hours earlier, a bulletin that no one in the Horning family but Tom seemed to have paid any attention to. They never heard an emergency siren. What they did hear was a sharp clanging sound coming from the basement. Judy, Lynne, and Lyle were still up, watching *Portland Wrestling*, and Bobbie was curled up with a book, keeping one eye on the teenagers, when they all heard what sounded like the metal garbage can bouncing down the basement stairs, making a terrible racket. Which made no sense, given that the garbage can was kept outside, next to the basement door. Maybe Whitey, the family's half-feral cat, had knocked it over—which still didn't make any sense. Everyone looked up at the sound. Then the lights all went out, the TV screen flickering to black. Everyone looked out the front window, their eyes drawn by the glimmer of moonlight on water, black water that now entirely surrounded the house, water where a yard and driveway should be, water already higher than the living room floor, or so it appeared.

"The boys!" Bobbie shouted, and she ran to the kitchen and out the back door and out onto the little porch, several steps above the yard. Water lapped just below the porch, an inky, swirling moat. To the north, a current was racing through a channel of water that now ran between the house and the cottage, flowing west toward the ocean like a fast-ebbing tide. "We have to get the boys!"

"Hold it!" Lyle was right behind Bobbie, and now he grabbed her—a strong woman, as tall as he—and held her to keep her from wading into the fast-moving water. "You can't go out there. Just wait. I'll go out and get them, soon as the water drops," he told her. "It's not getting any higher. The water's got to be dropping soon." She set her mouth, then nodded, irritated to be told what to do by a seventeen-year-old boy, but knowing he was right. Lyle let go, and the four of them stared at the little shingled cottage, a square, pale gray island in the stream, willing it to stay put and not get battered by logs or picked up and swept away with the debris they could see floating toward the bay mouth.

Gradually the pace of the water slowed. Gradually—it felt to Bobbie like an hour, but it was probably not more than fifteen minutes—the water level dropped and drained, and the river returned to its channel, and all that was left between the house and cottage was a wet, sandy, debris-strewn beach.

The Horning children wandered around in the moonlight for a half hour or more, not venturing much beyond their own yard, where there was plenty to see. The family's new blue Rambler station wagon, which Bobbie had bought not six months earlier, was ruined; the flood had picked it up from its spot in front of the house and floated it about a hundred feet. Now it was resting near the end of the driveway. Beyond it, across the road, the Finnish Meeting Hall was a shambles. It had apparently floated off its foundation, swung around 45 degrees, and then collapsed in a heap, surrounded by big driftwood logs. The meeting hall, their car, their own house and cottage— every surface within two or three feet of the ground was covered with big globs of cream-colored sea foam.

Sometime after midnight Bobbie finally gathered up the kids and herded them into the house. No water had seeped into their living area; at the height of the flood, the water hadn't actually risen much above the level of the floor, despite how it had appeared through the living room window. But the force of the wave had pushed open the basement door, which sat a couple of feet lower than the main floor of the house, and water had flooded down the stairs, carrying the metal garbage can with it and filling the basement with seawater that destroyed the furnace and hot water heater. Otherwise, the house seemed to be more or less intact, and the danger over. Bobbie lit a fire in the fireplace, gathered blankets and sleeping bags, and settled all the kids together in the living room to finish out the night, much as she did when winter storms knocked out the electricity.

At first light, Tom was back outside, checking on the pets. Muffin, of course, was fine; the dog had spent the night inside curled up with the kids on the living room floor. The chickens were indeed all alive and, judging from the murmured clucking coming from their coop, apparently none the worse for their short cruise from one side of the yard to the other. Handlebars, the French Lop rabbit that lived in the outside hutch, was also unharmed. But when Tom went inside the cottage to check on the residents of the hamsterium, it was another story. All the hamsters were dead. So was the indoor rabbit, which Tom found halfway under the steamer trunk, foaming at the mouth. All had apparently been electrocuted when the flood waters had come in contact with exposed electrical wires. Both the rabbit and the hamsters had developed a bad habit of gnawing the insulation off the Romex wiring Tom's grandfather had strung along the wall during the previous summer's remodel.

Whitey the cat was nowhere to be seen.

Tom poked around in the cottage some more. The sand now covering the floor was two and a half inches thick. A faint line of scum on the walls and lingering dampness on the lower stairs showed precisely how high the water had risen inside: two and a half feet, nearly as high as the doorknobs, almost to Tom's waist.

Back outside, Tom and Chris started picking up the ghost shrimp that littered the sand-covered yard—their waxy pink bodies flaccid, not designed to exist outside of their sand tunnels—and throwing them back into the river, where flocks of gulls now swarmed, scavenging. Hundreds of earthworms had wriggled up through the sand as well, joining the shrimp and flounders and sole and perch scattered there. The air was ripe with the smells of life churned up in the estuary. That, and something more. The septic tank, abandoned the previous summer when the city sewer line reached Venice Park, had floated up and out of the ground like a big balloon. Now it poked through the sand, the sulphurous gasses within it seeping out and mingling with the estuarine perfume of salt and decay.

Tom's and Chris's bikes, tucked in the old detached garage, were unharmed, so the two boys grabbed them and did a slow cruise of the neighborhood, riding on the sand that now covered the neighborhood's dirt lanes and asphalt streets, past shingled houses whose bathtub rings of yellow scum clearly delineated the previous night's high-water line. They saw cars on their sides up against houses and huge logs across driveways, blocking doors and windows. The Jacksons' house, two blocks from the Hornings', sat alone at the end of the point, where north-flowing Neawanna Creek curves west to join the Necanicum River. Even before they reached the house, they could see it was a disaster site. Nearly all the landscaping around the house was gone, and the living room's west-facing picture window looked as if a bomb had hit it; the actual culprit, a drift log five feet in diameter, lay on the sand-covered yard a few feet away. They peeked inside the waterlogged living room and could see the family's baby grand piano lying on its side in front of what had been the house's east wall but was now a gaping hole. A couple of blocks to the south, the boys stopped at the sand-covered highway and stared at what was left of the railroad bridge over the Neawanna: wooden supports dangling from iron rails.

Tom didn't venture much outside of his own neighborhood that day. It made him uneasy, somehow, all that chaos, things not being in their right places, not unlike the way he'd felt after his father died. What most interested him, anyway, was the river and the riverbank where he'd been digging the afternoon before. He headed there after lunch, half expecting to find his carefully excavated caves destroyed. But when he reached the end of Twenty-Sixth Avenue and dropped over the bank, there were the little garages he had dug, entirely intact. The tidal wave had clearly flooded his excavation site, but the riverbank and the walls of his hand-dug garages had held. That surge of water that had filled his own basement and blown open the Jacksons' house and torn up bridges had apparently flowed swiftly but sweetly past his excavation site, leaving almost no evidence of its passing.

The day was just as fair as the day before, mild and sunny. Tom plopped down on the bluff to watch the water for a while. The sand flats were bare, typical of low tide in the bay. Which made no sense; low tide should have been hours away, Tom knew. In fact, the bay should have been full of water by now, approaching high tide. Then he saw first one wave and then another roll up the river channels, waves rushing upstream cresting two or three feet high. It was the fastest incoming tide Tom had ever seen. He was mesmerized. This he had to see at close range. He began walking out across the sand flats toward the bay mouth.

He was about two hundred feet from the bank when he saw water begin to spill out of the channel and move toward him. Nothing dramatic: the water just crept across the sand flats in one long, low, relentless surge, covering them with an inch or two of dark, sand-laden water that spread ceaselessly, its leading edge jumping and popping, as if tripping over itself. Tom began walking backwards, watching and waiting for the pause between waves when the water would sink into the sand before the next wave lapped over it. But there was no pause. He picked up his pace. The water kept advancing, a dark, liquid creep. Finally he turned and scurried for the bank, running the rest of the way down Twenty-Sixth Avenue to his house.

When he got there, he found the rest of the family in the backyard, watching this new surge. Some of the debris that had floated into the estuary on the retreating wave the previous night—logs, appliances, pieces of the Fourth Avenue Bridge, which had been demolished by the tsunami—was now rolling and tumbling on the waves, gyrating as if the bay were boiling. It was like an encore, churning up more foam and filling but not quite spilling over the banks of the river and creek.

By now tourists had begun trickling through the neighborhood, rubbernecking the damaged houses and sand-layered lawns. There were always plenty of tourists in Seaside, but they generally stayed downtown or on the beach; Tom had never before seen a tourist in his neighborhood, north of downtown and well off the highway. He watched as car after car slowly cruised down Pine Street, pointing and smiling at Tom and his siblings through car windows. Some got out of their cars to take pictures of the damage or pluck souvenirs from wrack-strewn front yards, as if collecting shells on a beach. One car came to a slow stop in front of a house that had Japanese glass floats tucked into the rose garden—precious floats the owners had themselves found on the beach, floats that, surprisingly, hadn't drifted away but had stayed put, lodged in the mud, when the previous night's tidal wave had flooded the yard. A car door opened, and a boy jumped out. He ran across the yard, snatched one of the floats, and high-tailed it back to the car to deliver it to the smiling driver, apparently the kid's father. It was odd; they didn't look like crooks. They looked like any other tourists. But what Tom saw were thieves, people stealing from his neighbors. To the east, Tom saw black vinyl records flying over the treetops; someone must have entered the Jacksons' house and was filching, or just vandalizing, their record collection. The tidal wave seemed to have suspended ordinary rules of behavior.

Not until the next day, when the *Sunday Oregonian* newspaper arrived, did the Hornings and their neighbors get the big picture. The front page featured photos of the damage in Anchorage: exterior walls of the five-story J. C. Penney store sheared away, collapsed into a pile of rubble; stunned residents staring at the gaping hole where Fourth Avenue used to be, below a banner advertising an upcoming production of *Our Town*. President Johnson had declared the state of Alaska a major disaster area. With communication lines fractured, estimates of damage and casualties were hard to come by, but early reports suggested the death toll in Alaska could exceed fifty.

Miraculously, no one had drowned in Seaside or Cannon Beach, Oregon, or anywhere on the Washington or British Columbia coasts. The only disaster-related death in Seaside was that of Mary Eva Deis, the fifty-year-old sister and housekeeper of one of the priests at Our Lady of Victory Catholic Church. Apparently she and Father Deis had been in the process of evacuating by car when they stopped briefly in front of the fire station, at the very moment the chief of police sounded the siren directly overhead. The sudden wail amid the panic of evacuation must have been too much for Miss Deis's heart.

But down in Newport, one hundred miles to the south, four children camping with their parents on the beach had drowned when the arriving wave engulfed the driftwood shelter they were sleeping in, sparing the parents but sweeping all four children out to sea. Worst hit of all, outside of Alaska, was Crescent City, on the northern California coast. Eleven people had been killed and thirty blocks of homes and businesses flooded. Waves had flipped cars around, piling them up, driving them into and under buildings. Houses had been picked up and dropped on the highway. Some of the victims had been enjoying an evening out and had been trapped in waterfront restaurants and bars by the surge. A number of fishermen, upon hearing the belated warning, had begun motoring out of the harbor in an effort to evacuate to deep water, but the gathering wave had sucked so much water out of the harbor that it grounded their boats and forced the fleeing fishermen to dash on foot across the exposed harbor bottom. Another boat on the bay had overturned, tossing seven people into the churning water. One man clung to his house as he watched the water carry his wife away. Homeless residents huddled at the county fairgrounds where families had hunted Easter eggs the day before. South of Crescent City the tidal wave lost most of its punch; it wasn't big enough to do any damage by the time it reached San Francisco and San Diego, where thousands of people had flocked to the beaches and shoreline viewpoints to see it, ignoring pleas by police to get to high ground. Tide gauges as far as Japan and Chile and even the Palmer Peninsula in Antarctica registered the waves' arrival.

There was a silver lining, for Seaside, at least: scientists had no reason to believe that a tsunami had ever before struck the Oregon coast, nor had they any reason to think that another would strike again any time soon. "Wall of Water First for State," declared the headline in the *Sunday Oregonian*. "The U.S. Coast Guard reported Sunday night that its records showed no previous tidal waves in Oregon or Washington. Threats? Yes. Tidal waves? No."

As residents of Venice Park began the job of cleaning up—ripping out waterlogged carpets and replacing furnaces and hosing sand off floors—the Hornings and their neighbors began to realize how lucky they'd been. Or how unlucky, depending on how you looked at it. If the tsunami had struck Seaside three or four hours earlier, at low tide, Tom's neighborhood wouldn't have flooded at all. The tidal wave might have manifested as nothing more than a series of surges up the main channel of the Necanicum, a cresting bore similar to the one that muscled up the river after the massive 1960 earthquake in Chile, busting up a few docks. They might not have called it a tidal wave at all.

If, on the other hand, the weather had been stormy, as it often is in March, the story might have been quite different. Had the surf been a few feet higher than the mild, two- to three-foot surf that Good Friday evening, waves wouldn't have merely splashed over the oceanfront Promenade and leaked through its vents. Waves five or six or even eight feet high would have crashed onto the Prom, sending logs through the houses all along on the oceanfront, the way the log did crash through the Jacksons' front window, washing torrents of seawater down every street downtown. Water may have risen not just two or three feet in Tom's neighborhood but four or five feet or more, filling dozens of houses to the ceiling as it had the Jackson's house and just a few others at the bend in the Neawanna. Had just one or two factors been different, the hospital's ground floor, which had been dampened by the wave, and perhaps a school or two would have been inundated. A lot of people would likely have drowned. A couple more feet of water, and the little cottage Tom and Chris had been sleeping in would certainly have been wrenched free from the pilings it was anchored to, would have been swept up in the tsunami's retreat and floated out to sea.

Could Oregon have an earthquake of the size that had struck Alaska? It's unlikely anyone in Seaside even dwelt on that possibility. Alaska was rife with earthquakes; in Oregon, earthquakes were rare. And Seasiders were far too busy pointing fingers about the last tsunami from afar to worry about an earthquake closer to home. "Something must be done to provide authentic warnings on tidal waves," read an editorial two weeks after the fact.

The cleanup was the most immediate concern, especially for residents of the inundated neighborhoods. Debris lay strewn throughout Venice Park and between the oceanfront houses lining the Prom: drift logs, appliances, propane tanks, bridge parts, lawn furniture. Thick foam clung to the lower walls of buildings, begging to be scrubbed, and a layer of ocean sand had settled on lawns and roads, everywhere the tsunami had reached before withdrawing. Most of the bridges crossing the Necanicum River had been seriously damaged. The old wooden Fourth Avenue Bridge had been razed by the surge, which played itself out at the Seaside golf course, collapsing in a scattering of debris across the fairways and leaving a bathtub ring of buttery scum around the gently sloping basin rising from the bunker at the ninth green.

The owners of the Twelfth Avenue Grocery lent Bobbie their delivery car to use for a couple of weeks until she was able to buy a new one. Members of the Hornings' church pitched in to help with the cleanup at Tom's house, shoveling out the basement and washing the sand out the door of the cottage and scrubbing its floors. While the adults worked, Tom played outside, using a hose to carve moats and gullies and canals in the layer of tsunami sand covering the lawn.

Whitey the cat never reappeared. Tom had a hard time believing he'd drowned. He liked to think that Whitey had been swept upstream in the surge and had found a new house to live under, a new family to adopt. Whitey was a tough old boy.

At least it was over. In an editorial immediately following the tsunami, the *Seaside Signal* captured the sense of relief that dominated the community's mood following the only such waves ever known to have struck Seaside, causing the biggest disaster in local memory. Clearly someone had "slipped badly" in failing to warn Seaside and other Northwest coastal communities of the approaching tsunami, the editorial writer acknowledged, but there was no need to be overly concerned about the future. "This thing will probably never happen again in our lifetime."

2

Barnacles Never Lie

PRINCE WILLIAM SOUND, ALASKA, 1964

*In questions of science, the authority of a thousand is not worth
the humble reasoning of a single individual.*

—Galileo Galilei

PEERING OUT THE AMPHIBIAN AIRCRAFT'S PASSENGER WINDOW, George
Plafker watched as the green claw of Evans Island, one of a cluster of long
islands nestled together in the southwest entrance to Alaska's Prince William
Sound, came into view. The day was faultless: blue sky, little wind, perfect
flying weather. Landing the Grumman Widgeon on the smooth waters of
Chenega Cove should be easy. March snow still blanketed the high peaks
that rose sharply behind the native village of Chenega, which was arrayed
on a narrow strip of beach between the dark, steeply ascending forest and
the sparkling bay. Then, approaching closer, the pilot banked, and Plafker
pointed his Olympus camera out the window. The scene that now filled his
viewfinder was chilling.

There was hardly anything left of the village he had come to know well
from previous visits while mapping the geology of the Alaskan coastline—
nothing left but the little wooden schoolhouse on a rise hard against the
forest, its flagpole standing flagless on the sloping ground. Off to the left, a
cluster of rude cabins sat shoved into one another among the trees. There
were some smooth patches between the schoolhouse and the beach—
foundations, apparently, of the homes that had stood there not two days
before. Now they were gone. The people here hadn't even had time to run;
while at other locations the waves followed the quake by fifteen or twenty
minutes, here, a wave had enveloped the beach even before the shaking
had stopped. From what he'd heard, back at the air force base outside of
Anchorage, the survivors had been relocated to the village of Tatitlek, on the
sound's northeastern shore.

Sun had melted the snow off the spruce trees, but the ground was still white—everywhere but the shore, where scallops of snow-free sand and rock and low shrubs stood out sharply: a tracing of the upper limits of the wave, Plafker realized. The highest scallop was that just below the schoolhouse, at the head of the cove; the killing wave had apparently focused its energy here. On any other day, the head of the cove would have seemed the most protected spot of all.

Plafker, a staff geologist with the U.S. Geological Survey in Menlo Park, California, was still reeling a bit from the circumstances that had landed him, with less than twenty-four hours' notice, in Alaska the day before. The trip was a young geologist's dream. On the evening of March 27, 1964, he had been in Seattle, there to deliver a paper on sedimentology at the western section meeting of the Geological Society of America, when news began circulating about a big earthquake in Alaska; it was all over the radio and television news. Earlier that evening some geologists attached to the meeting had gone up the Space Needle, still a novelty since its opening two years earlier at the 1962 Seattle World's Fair. Only later did they realize what had caused the slender 605-foot tower—1,400 miles south of the Alaskan earthquake's epicenter—to unexpectedly sway.

IT WOULD BE A DECADE BEFORE SCIENTISTS, using a new moment magnitude scale developed to measure earthquakes' relative strength, would set the figure for this quake that struck Alaska's Prince William Sound at 5:36 p.m. on March 27, 1964, at 9.2—the largest earthquake ever recorded in North America, the second-largest quake recorded anywhere in the world, behind only the magnitude 9.5 quake that had erupted off the coast of Chile just four years earlier. It triggered waves in lakes as far away as Texas, stirred water in wells in South Africa, on the other side of the Earth. The epicenter was seventy-four miles southeast of Anchorage, the state's largest city, home then to about one hundred thousand people. The intense shaking collapsed buildings and bridges, tore apart roads, snapped off the tops of trees, and triggered landslides, damaging or destroying about thirty blocks of homes and other buildings in Anchorage alone.

Most of the property loss—eventually estimated at more than $300 million in 1964 dollars—was caused by buildings collapsing from the violence of the quake. Ten people died from injuries suffered during the quake itself. But it was the tsunami waves following the quake that caused the greatest suffering. Surging waters ravaged the Alaska shoreline south of Anchorage with huge waves, waves thirty and forty and one hundred and—in Valdez Inlet—nearly two hundred feet high, higher than the tallest building in the entire state, that inundated the low-lying parts of the coast with unstoppable walls of water. The tsunami surges associated with the earthquake not only damaged buildings but wiped entire communities off the map, killing 122 people—more than ten times as many as were killed by the quake itself.

In Whittier, an army port town and railroad installation about halfway between the earthquake's epicenter and Anchorage, the shaking had lasted some six minutes. Sea waves began to engulf the town even before the ground stopped moving. When the first small wave hit, railroad maintenance man Jerry Ware hopped in his car and sped to his house trailer, hoping to get his wife and six-month-old daughter out of the trailer and up to higher ground before another wave struck. He arrived just ahead of the forty-foot-tall, debris-laden second wave. It smashed the trailer, impaling Mrs. Ware with pieces of wood and sweeping the baby out of her arms. The couple survived that wave and a third—this one thirty feet tall—by climbing onto a railroad freight car. Rescuers combing the high ground later found their baby daughter alive on a snow bank, where the wave had deposited her, though she died soon after. Twelve other Whittier residents were swept out to sea; their bodies were never found.

To the south, at the port town of Seward on the Kenai Peninsula, the earthquake caused a massive oil tank on the waterfront to collapse and explode. A section of the waterfront slid into Resurrection Bay, creating an immediate tsunami that smashed the shoreline and spread burning oil and debris. Twenty minutes later, a second, larger tsunami wave struck, demolishing railroad cars, houses, airplanes, and airplane hangars, and an hour after that an even larger wave roared in, descending on the now-dark harbor as a thirty-foot wall of flame. Residents escaped by scrambling into trees and onto rooftops; some who tried heading to the hills were thwarted by bridges that had collapsed during the earthquake. Twelve people died. Only the day before, residents had been celebrating the news that Seward had been selected as an All-America City.

In Kodiak, two hundred miles to the southwest, earthquake damage had been minimal; some buildings had been knocked off their foundations, and the high school had been slightly damaged, but no one had been killed. When, a half hour later, the emergency siren at the fire station began to wail, a few people in the downtown waterfront area headed uphill, while other residents, overcome by curiosity, headed down to the port to see what was going on. The local civil defense director, hearing the siren, assumed a house was on fire. The first surge flooded the lower part of the downtown business district with hip-deep water, but the second wave was the killer. It gushed over the breakwater, elevating the boats moored there, and for a few moments they seemed to hover higher than the town, floating at the crest of the wave. Then every boat in what was then Alaska's largest king crab fishing center was swept inland several blocks and driven through buildings or dumped on what was left of downtown—which wasn't much by the time the wave withdrew. Buildings had been rearranged, tipped over, smashed by the surging water. Canneries along the waterfront were destroyed. Eight people were killed, all of them either aboard boats in the harbor or attempting to board boats, hoping to steer them out of harm's way.

Similar scenes were played out in Cordova and Seldovia and other communities, but the greatest number of casualties was in Valdez, located at the head of a fjord fifty miles east of the earthquake's epicenter. In the span of ten minutes, thirty-one people died as the town's waterfront virtually vanished under a series of towering waves. Twenty-eight children and adults had been standing on the city dock watching the unloading of the coastal supply freighter *Chena*, a big event in this isolated port town. All twenty-eight died, along with two longshoremen aboard the *Chena*, when the dock was engulfed by a wave estimated at more than one hundred and fifty feet tall. All but two of the seventy boats in the harbor were destroyed. Shortly before the quake struck, local resident Harry Henderson had left the harbor in his boat, motoring toward his bayside cabin down the shore. Harry, his boat, and his cabin were never seen again.

But no community suffered a greater blow than Chenega, a native village of seventy-six people on a remote island at the edge of Prince William Sound. Chenega was arrayed on a narrow curve of coastal plain at the base of steep mountains, which were still almost entirely covered with snow in late March. Most buildings in the village were clustered close to the beach— all but the school, which was built on a knoll a short walk above the town. When the ground began to shake, many of the seventy or so people in the village that evening began to walk toward the white clapboard Russian

Orthodox church, the community's focal point and the biggest building in the village. But the tsunami arrived even before the shaking stopped. In an instant, a seventy-foot-tall wave swallowed one-third of the town's residents, most of their houses, and the church. Those who got to higher ground and were spared—the parents and grandparents and children and cousins and lifelong friends of the dozens just drowned—huddled high on the treeless shore for hours, watching as night fell and more big waves surged onto the beach, watching their possessions and the remains of their homes drift out to sea: now they were nothing but black debris, smeared with moonlight and bobbing above the surface of the dark, turbulent sea.

It could have been worse. The tsunami had struck Prince William Sound, an area famous for huge tidal fluctuations, at low tide. And not many people lived in the southern Alaska coastal towns struck by the waves; the state's entire population in 1964 was just two hundred fifty thousand.

The tsunami then continued south, fanning out across the Pacific and marching down the mostly uninhabited outer coast of the Alaska panhandle and British Columbia. Moving through the deep ocean faster than a jet airplane could fly the same distance, it took less than four hours to travel some thirteen hundred miles from the earthquake's epicenter near Anchorage to the northern tip of Vancouver Island. What appeared on the open ocean's surface as only a low hump of water a couple of feet high slowed and then mounted higher as it approached the shore, finally cresting as a large wave that barreled up narrow Quatsino Sound to the town of Port Alice, where it tore up boat ramps and seaplane moorings and scattered cut timber on the waterfront like Tinker Toys. All along the island's west coast, the tsunami pulled buildings off their foundations, sank boats, and broke apart log booms. It surged into wide Barkley Sound, chasing dozens of startled teenagers off the beach, then began funneling up thirty-five-mile-long Alberni Inlet, strengthening as the narrow fjord concentrated its force, until it collapsed on the twin towns of Alberni and Port Alberni, sweeping log booms and boats onto roads and cars into fields, and lifting houses—from tourist cabins to two-story homes—off their foundations.

Southward the wave continued, past Cape Flattery at the northwestern corner of Washington to the Olympic National Park resort community of Kalaloch, where two eleven-year-old children camping on the beach were rescued seconds before the debris-laden wave struck. At Moclips it crumpled the seawall and several houses, washing one homeowner out of his house, onto a road, and nearly back out to sea. It picked up a trailer parked on the beach south of the Copalis River, tumbling it and spitting out its occupants—a couple and their three children, ages seven to twenty—who suddenly found themselves swimming in the dark, dodging logs. They managed to reach shore—bruised, chilled, and stripped of most of their clothes, but alive. The wave surged into Willapa Bay, just north of the Columbia River, and up the rivers that feed the bay, damaging bridges and breaking up vast rafts of logs headed to sawmills, scattering timber on the shoreline and across the churning surface of the bay.

It took just a minute or two for the tsunami—greatly diminished by now, but still packing a punch—to travel from the mouth of the Columbia River to the town of Seaside, seventeen miles to the south, and less than an hour for it to go the length of the Oregon coast and crash into the harbor at Crescent City, California.

As soon as he heard the news, Plafker and his colleagues were on the phone with USGS headquarters; clearly someone from the Survey needed to go up to Alaska right away to take a look around. Plafker was a logical choice. Thirty-five years old, he'd spent five years as a petroleum geologist with Chevron and seven years with the Survey, scrutinizing, among other places, the Alaska coastline, so he knew the territory well. Plus, he was already halfway to Alaska from his office at USGS headquarters in Menlo Park, California. Other than a little secondhand fieldwork on the San Andreas Fault in California, he had never done any post-earthquake studies—hardly anyone had. But, as was often the case in the Survey back then, you were expected to show up, rely on your training, and figure it out along the way. He would be accompanied by two others from the USGS. Arthur Grantz, also a geologist, was attending the Seattle meeting as well; he had already done fieldwork in the Matanuska Valley north of Anchorage and would be a good choice for examining post-quake landscape changes in the state's interior. They'd be joined by Reuben Kachidoorian, an engineering geologist who could look over structural failures such as broken bridges and cracked roads. Kachidoorian was still down in Menlo Park; he would stop by Art's and George's houses and grab extra clothing and gear for them before catching a morning flight from San Francisco to Seattle, in time for the three of them to board the mid-day flight to Anchorage.

The earthquake the day before had collapsed the six-story control tower at Anchorage International, killing the air traffic controller on duty, so their plane was diverted to Elmendorf Air Force Base on the outskirts of town. The diversion was fortuitous for the visiting geologists. At Elmendorf they had government housing at their disposal, and the air force could equip them with Alaska-caliber winter coats and boots they could wear in the field during their weeklong reconnaissance. Ground travel outside the Anchorage area was impossible—bridges and roads were damaged and destroyed all along the coast and even inland—so the air force gave the team the use of its helicopters, and the commanding general at the base even lent them his personal twin-engine Beechcraft for a visit to Kodiak. The Widgeon, chartered from Cordova Airlines to access remote parts of Prince William Sound, and its pilot, Jim Osborne, were both old friends of Plafker's; he'd often flown with Osborne to reach the state's more remote lakes, estuaries, and coastlines during previous fieldwork.

By 1964, it was understood that earthquakes tended to cluster in certain areas of the Earth. But geologists weren't yet clear about what mechanism actually caused the Earth to occasionally shake violently. "There is some evidence that full moon, high tides, heavy rainfall, sharp changes in barometric pressure, and, especially, another earthquake elsewhere act as 'triggers' for earthquakes," reads *Geology: Principles and Processes*, a leading 1960 textbook. "However, detailed studies by seismologists, especially in quake-ridden Japan, have failed so far to make useful predictions possible."

Even less well understood were what were commonly called tidal or seismic sea waves. These fast-moving, far-ranging, highly destructive waves were often associated with earthquakes, though the exact nature of that association wasn't clear. Rarely was such a wave referred to as a "tsunami"; scientists had only recently borrowed the term from the Japanese language. (Its literal translation, "harbor wave," captures the way a tsunami's power is magnified inside bays, in contrast with storm waves, which may break on the beach but tend to flatten out within a harbor's confines.) *Principles and Processes* specified that seismic sea waves were caused by *earth*quakes— seismic disturbances on dry land—and were not to be confused with "seaquakes," which it defined as tremors of the water resulting from underwater quaking. These seaquakes, it said, "merely shake ships, stun and kill fish, and ripple the surface of the sea."

Plafker, heading to Alaska mere hours after a massive quake and tsunami, hadn't really known what to expect, and he wasn't sure what to look for other than the obvious—the off-kilter look of land that had uplifted or sunk or lost strength or slid, and the swept-clean look of shorelines that had been washed by waves of unprecedented height. The real action had happened underwater, the apparent location of the fault where the quake occurred, increasing the challenge. But Plafker was familiar enough with the country to notice changes above the water that might offer clues even to what had happened on the seafloor. The team got straight to work. There were ample examples of ground failure—places where seemingly solid earth had turned to liquid, collapsing whatever had stood on it. Whole hillsides had slumped and slid. They measured how far up the tsunami had run in a variety of locations, noting what direction the waves had come from, and made inquiries about the time span between the shaking and the start of inundation. The tsunami questions were particularly tricky; some of the first surges and the highest measured wave run-ups were apparently caused by underwater landslides in fjords or bays, slides triggered by the shaking, while other, later surges seemed to have originated out in the ocean.

What most struck Plafker, however, were reports he began hearing about water level changes since the quake, reports that the tides were noticeably higher, or lower, than they used to be—in fact, in very few places he visited did the water level seem to be the same as it had been prior to the quake. Clearly the sea level itself hadn't changed, so it must be the height of the land that had shifted up or down. It was tough to quantify the changes, however. There weren't a lot of tide gauges deployed at the time, so pre-quake benchmarks were slim. More helpful were Alaskans themselves, people who lived by and from the sea and who knew it intimately. But much of the affected shoreline was uninhabited, and people's recollections after a disaster aren't always reliable.

Rather, Plafker's best, most reliable informants turned out to be some coastal natives that had survived the quake and tsunami and were still there, clinging to the shoreline of that rough, cold country; they had nowhere else to go. They were the acorn barnacles, *Semibalanus balanoides*, found by the millions everywhere the shoreline was rocky rather than sandy—which was almost everywhere. Acorn barnacles live inside six-walled calcareous shells shaped like little volcanoes with a craterlike opening at the top and an *operculum*, or lid, shaped like a little bird's beak. At low tide, when the barnacles are exposed to air, the operculum pinches closed to keep the animal inside moist. Acorn barnacles are ubiquitous in the tidal zone of the rocky bayshores and outer coast of south-central Alaska. These vast, bumpy colonies, Plafker realized, created a reliable elevation line set by the rising tide: the top of what marine biologists called the midlittoral or upper intertidal zone—roughly, the mean high-water line. Young barnacles, in particular, cannot survive prolonged exposure to air; their need to stick close to the waterline strictly limits the height to which an entire barnacle colony can rise. The olive-green rockweed common to these shores and the encrusting algae that spreads like a hard, gray film on the rocks in the splash zone just above the barnacles were helpful guides to land-level change as well, but their vigor and breadth are influenced by exposure to sunlight as well as by the tides. It was barnacles that offered Plafker the most reliable narrative about where, during the earthquake, the shoreline had abruptly risen and where it had plunged down. Barnacles never lie.

The March reconnaissance visit lasted only a week; it was hard to see much with snow still on the ground and the state still in chaos, and there was just so much to see, and so much ground to cover. In less than two months, Plafker and a much more robust team of researchers from the USGS would be back to spend the whole summer combing south-central Alaska to collect data.

But even in that first short visit, the geologists could already see a pattern developing. In some places—northern Kodiak Island, the Kenai Peninsula, and the fjords reaching northwest from Prince Williams Sound—the barnacle aggregations had been drowned and were now entirely submerged, even at low tide. In other places—elsewhere in Prince William Sound and in the Copper River Delta to the east, for instance—whole barnacle colonies were now high and dry, as much as seven feet above mean high tide. Clearly the shoreline had subsided in some places and in others had shot up to a new normal. Those land-level changes suggested alterations in the seabed as well, changes Plafker couldn't see but could infer.

It was pilot Jim Osborne who led Plafker to the most radical deformation site of all on the very first day of the March reconnaissance trip. Narrow, uninhabited Montague Island lies at the entrance to Prince William Sound. Osborne, flying past the island, had noticed a new band of white appearing on the mountain front defining the island's southwestern tip: the barnacle line, Plafker recognized, a line so thick and now so high above the waterline that a bush pilot could spot it from cruising altitude. There was no way to safely land the Widgeon on the surrounding waters that March day; as the island had risen, so had formerly submerged rocks and reefs. Plafker would discover an astounding thirty-eight feet of uplift on that side of the island when he visited the island by boat that summer to take measurements. Later Osborne would also point Plafker toward a band of landslides along the mountain front on the west side of the island that—reached by helicopter that summer—would turn out to be on the trace of a newly ruptured secondary fault.

The Montague Island fault was a substantial fault, but Plafker understood it was not *the* fault. The earthquake, the landscape ups and downs, and the main tsunami had all been caused by movement along what would come to be called the Aleutian Megathrust, a long fault that runs undersea for 2,230 miles from near the Kamchatka Peninsula in eastern Siberia eastward to the Gulf of Alaska.

Plafker's conclusions, initially summarized in an article published in *Science* the following year, would challenge conventional thinking not only about this earthquake and Alaska's geology but about similar geological settings worldwide. He proposed that the quake itself and the altered landscape (and, inferentially, altered seafloor) had been caused by the slow, progressive movement of the oceanic crust moving toward and pushing or thrusting under the continental margin. Plafker concluded that the earthquake had been triggered when the edge of the continent shifted at least sixty-five feet toward the sea—a huge leap. What had caused some parts of the region to subside and submerge and others—Montague Island, for instance—to rise was the elastic extension and thinning of the Earth's crust as a result of that fault line rupture, the way a block of foam rubber stretches and thins when pulled apart.

The barnacle line wasn't the only indicator of land-level changes; other members of the USGS reconnaissance team documented shoreline forests whose roots were now drowned in seawater as a result, presumably, of subsidence. These land-level changes, along with the quake itself, were all evidence of subduction of the Pacific tectonic plate under the North American plate, Plafker concluded, though not in those words; it would be a few years yet before *tectonic plate* and *subduction* entered geology's lexicon.

What's more, Plafker said, this earthquake clearly indicated that "vertical displacement of the sea bottom"—that huge leap, and the up and down movement of the seafloor that came with it—had caused the tsunami. Never mind that the epicenter—the point on the surface of the Earth under which the earthquake had begun—was under the continent; the quake had occurred along a long *line* that ran out under the sea, along the continental shelf and slope. Its displacement under the ocean was what had generated the main tsunami, complicated by secondary tsunamis from earthquake-triggered underwater landslides.

Plafker knew his paper was going to ruffle some feathers. Oceanic crust thrusting under the continental margin? That sounded suspiciously like an argument for continental drift: the old and much-maligned theory that, as the textbook *Principles and Processes* put it, the continents and oceans had "exchanged places because of warping or folding of the earth's crust." Don't believe it, the textbook's authors cautioned readers.

"Most geologists and geophysicists have long been convinced that the continental masses and the deep ocean basin have remained in the same general locations throughout recorded geologic time." That opinion was echoed in *The Earth Sciences*, another geology textbook of the day: "Drifting of a rigid continental plate across the basaltic oceanic layers," it stated, "is a physical impossibility."

In fact, just months before Plafker's paper came out in *Science* in the summer of 1965, the same journal had published a paper by the leading seismologist at Caltech, Frank Press—co-founder of Lamont Geological Observatory at Columbia University and eventually President Jimmy Carter's science advisor—who had come to a conclusion diametrically opposed to Plafker's about the origin of the Alaska quake. That led to some awkward moments at conferences—scientists jumping up and shouting at Plafker, "It's not in the data!" "It's physically impossible!" Who are you going to put your money on, a world-famous academic or an unknown USGS staff geologist? It would be a few years—1969—before Press would encounter Plafker at a conference and congratulate him for besting him on the Alaska earthquake story.

But the criticism hadn't been personal; Plafker knew it was all part of the scientific process. And he knew what he knew. Seismologists' conclusions were only as good as their instruments, and in 1964, the instruments used to study seismicity in the oceanic crust were still pretty crude. Meanwhile, Plafker had been to Alaska. He'd seen the drowned barnacles and the desiccated barnacles and the Montague Island scarp, its barnacle line as tall as four-story building. So his conclusions challenged the established thinking about this stuff, so what? Wasn't that what science was all about?

Anyway, Plafker had always thought continental drift made sense, even if the details hadn't quite been worked out and the theory was practically bringing some American and British scientists to fisticuffs. Between his stints with the USGS, Plafker had spent six years in Latin America prospecting for Chevron. He'd seen what had to be glacial till—*tillite*, actually, a kind of bouldery sedimentary rock composed in part of gravel scraped off rock by moving glaciers—on the east side of the Andes in Bolivia. No glaciers in sight there. Where could this till have come from, scientists there wondered. Glaciers, obviously, Plafker figured: probably back 350 million years ago, when South America was still connected to Antarctica, down around the South Pole.

Plafker had worked in Alaska. He knew what glacial till looked like. Till doesn't lie. No more than barnacles do.

3

Silver Dollar

TOM, 1964

ONE DAY IN LATE MAY, barely two months after the tsunami, Tom Horning came home from school to find his grandparents and mother sitting on a drift log in the yard under a tall spruce, watching the bay. Grandma and Grandpa Baker lived in Berkeley, California. They hadn't been up to visit since the previous summer, when Grandpa had taken the Hornings' little seaside cottage and transformed it into the kids' playhouse and Chris's and Tom's bedroom. Tom spotted him from the driveway as he rode up on his bike. His mother was usually working at that time of day; she must have left work to meet them. He ditched the bike in the garage, then eased around the corner, watching, overcome by a sudden shyness. It had been a year, after all. At the sound of Tom's feet on the gravel driveway, Grandpa, in his crisp white shirt and bolo tie, looked up, turning his head toward Tom, waving him over with a big smile, huge creases forming at the sides of the old man's tanned face. Tom approached slowly, and when he got there, Grandpa, ever formal, ever affectionate, reached out a hand and ruffled the hair on the top of Tom's head. At that touch, Tom knew all was right with the world.

Grandfather Baker was a tall man, taller even than his nearly six-foot-tall daughter, his white hair neatly trimmed, and his skin ruddy from the California sun. Grandmother Baker wasn't as tall, nor quite as tender, and Tom had developed a certain caution around her. Often Tom would see just his mother and grandfather sitting on that log side-by-side with their matching square jaws, she with her signature sweatshirt and cheap tennis shoes and straight back, he starched and just beginning to stoop. In him, you could see where Bobbie came from, just as, looking at Bobbie, you could see the man Tom would become: tall and big-boned, a little quirky, a square peg. Grandfather didn't busy himself with a lot of chores this visit, as he had during his visit the previous summer. He and Grandma Baker were on vacation, heading north to catch the ferry up the Inland Passage to Alaska and Canada. Most days he seemed content to just sit with Bobbie, to watch the gulls circling over the estuary, the osprey diving for fish, the flocks of sanderlings that wheeled and spun like a gust of mirrors. Sometimes Tom would take walks with his grandfather, the older man always teaching, pointing out a little spruce seedling taking root on a drift stump at the edge of the bay, cocking his head to hear the call of a songbird hidden in the branches of a shore pine.

Born in 1890, Fredrick Storrs Baker was a forester, dean emeritus of the University of California School of Forestry, author of a leading silviculture textbook, a professor popular with students for his enthusiasm and utter lack of pretense. It was he who had taught Bobbie how to skip rocks across the surface of a creek. His parents had been amateur botanists, so forestry had been a natural career choice, just as a career in entomology had seemed natural to his daughter Bobbie, a tall tomboy with an aptitude for science. She could name all the trees and wildflowers but she loved even more to watch a spider weave a web outside a window, to follow a beetle scuttling through the duff on the forest floor. Ultimately she got bogged down trying to memorize the minutiae of insect morphology: the plumose and serrate and pectinate antennae, the hundreds of different vein patterns in the forewings and hindwings. Bobbie had always been more interested in books and bugs than in boys; as a young woman, she was taller than most men, favored practicality over fashion, was quick with a laugh but never coy. Her own mother had suggested, and she'd agreed, that with her looks and manner, marriage was probably not in her future. Nursing would require less schooling than entomology and would guarantee self-sufficiency. So Bobbie went to nursing school, then promptly—to her mother's surprise, and perhaps to her own—had married John Horning, the son of one of her father's friends and forestry colleagues, and moved north to Oregon.

Two Christmases before, Grandpa Baker had given Tom a silver dollar, dated 1889. The palm-sized coin had been heavy and cold in Tom's hand, and weighted with promise, and with mystery. One of his friends at school knew something about coins and suggested Tom was probably a millionaire now, or at least a *thousand*aire, making him something of a minor celebrity at Central School for a few weeks, until they realized the coin wasn't a rare dollar from the Carson City Mint but a more pedestrian—albeit old—coin minted in San Francisco.

But it wasn't the coin's monetary worth that Tom found compelling. What are silver dollars made from, he wondered. Mostly silver, his grandfather had told him, but with a little copper too. So where did that copper and silver come from? Tom knew about silver mines, about silver being dug out of the earth. But what did it look like when it was in the earth, as raw rock; was it shiny like the coin, or dull as the gravel in his driveway? How did you recognize silver in the earth, and how did you turn it from something earthy into something precious? And what forces created silver in one mountain and copper in another? Where did it all come from?

Those questions always led to a question that had simmered in the back of Tom's mind for as long as he could remember: *Where did I come from?* It wasn't the *how* but the *who* that worried him. His father had died when he was just five years old and was rarely mentioned; maybe he had never existed. His grandparents all lived far away; Tom had begun to wonder if they really were his grandparents.

Except for Grandpa Baker. When he was around, Tom felt he had a place, was part of something—fatherless, perhaps, but not grandfatherless.

But this would be Grandpa Baker's last visit to Seaside. He died just seven months later, on New Years Day 1965, killed by congestive heart failure that must have been taking its toll even as he sat reminiscing with Tom's mother on the drift log, as he and Grandma Baker drove out the driveway, waving, headed north to Alaska. Some family members later speculated that it was seeing the tsunami debris—the logs in the yard, the broken patio wall—that had triggered his precipitous decline. Thinking back on it, Tom had to concede that the visit probably was hard on him: seeing his daughter, still recovering from surgery for breast cancer, raising five lively kids by herself, and on top of that, a tsunami that had wrecked her car and her furnace and nearly swept away two of the kids.

But Tom was pretty sure his grandfather had been as thrilled by the tsunami as he himself had been. An earthquake in Alaska rearranging the furniture thousands of miles away in Seaside, Oregon; nature asserting itself in unexpected ways; science challenged to find new answers to questions as old as time: how, Tom wondered, could that have failed to excite Grandpa Baker?

4

A Flexible Worldview

LONDON, ENGLAND, 1964

Many [scientists] don't take readily to new ideas, and others, a bit too readily.

—Sir Harold Jeffreys

EXACTLY ONE WEEK BEFORE ALASKA'S Good Friday earthquake, an esteemed group of geologists, physicists, geophysicists, and engineers from around the world had gathered inside the Lecture Theatre of the Royal Institution, a palatial clubhouse for scientists on London's Albemarle Street just a short walk from Hyde Park. The meeting was organized by Britain's Royal Society, the world's oldest science academy. Among those in attendance were many of the best minds from some of the world's leading research institutions: Oxford and Cambridge, Scripps Institution of Oceanography, Caltech and MIT, NASA's Goddard Space Flight Center, the University of Toronto, the University of Utrecht, Service de la Carte Géologique de la France.

The purpose of the two-day symposium, one of many such symposia that would take place around the world over the next few years, was to reconsider a theory that had been formally proposed a half-century earlier but ultimately dismissed by almost every geologist in the world as unproven at best or, at worst, ridiculous: the theory of continental drift. By the end of the decade, it would have a new name, one that better reflected a new understanding of how the continents and ocean basins moved: on massive plates that collectively engaged the Earth's crust and upper mantle, or lithosphere, in a continuous process of building—*tektonikós*, in ancient Greek—and rebuilding the Earth. A revolution in the earth sciences had begun, one that would ultimately require a complete revision of every geology textbook in the world. New research, much of it funded by the American military during and after World War II, had begun to cast continental drift in a far more favorable light.

45

MISSING FROM THAT 1964 GATHERING in London was the one individual most closely associated with continental drift, the theory's most ardent fan and tireless proponent, German meteorologist and Arctic explorer Alfred Wegener. Born in 1880 and raised in Berlin, Wegener was an avid hiker, climber, and recreational hot-air balloonist in his youth. In school he gravitated toward the natural sciences. He completed a PhD in astronomy in 1904, then shifted his career interests to meteorology, taking a job as a technical aide at an aeronautic observatory outside of Berlin. At age twenty-five, he was invited to join a two-year Danish research expedition to little-explored Greenland. He became the first person in the world to collect high-altitude weather data in a polar climate using kites and balloons.

Soon after his return, Wegener took a position as a lecturer at the university in Marburg, teaching meteorology and astronomy but equally intrigued by scientific puzzles outside his own sphere of expertise, such as the undeniable congruence between the continental coastlines on either side of the Atlantic. Could the continents once have been connected, as had been proposed over the centuries by other thinkers? Not likely, he decided. But a year later he happened upon some fossil evidence purporting to prove the existence, millions of years earlier, of a land bridge of some kind between Africa and Brazil. What if it hadn't been a land bridge? What if the two continents had actually been part of one bigger continent? What if they had somehow broken up and "drifted" apart—like the chunks of ice he had seen calving from Greenland glaciers, their ragged edges matching their mother glacier's edge as they floated away across a widening chasm of black ocean? Wegener started digging around in geological and paleontological literature, summarizing his thoughts in a 1912 paper entitled "The Geophysical Basis of the Evolution of the Large-Scale Features of the Earth's Crust."

There was, at that time, broad agreement among scientists that the mountain ranges and other features of the Earth's surface hadn't sprung into existence fully formed but had evolved over time. The question was, how? Discovery of fossilized tropical plants and animals in temperate and even subarctic locations where such species could not conceivably survive today demanded explanation, as did the discovery of identical fossilized animal remains on continents separated by thousands of miles of ocean—distances far too vast for such animals to swim or fly. In Wegener's time, two theories prevailed. One held that the Earth had for millennia been cooling and shrinking and—much like a baked potato after it comes out of the oven, or an apple drying on a shelf—was contracting and wrinkling in the process. A companion theory suggested that land bridges that had long ago sunk beneath the ocean had once allowed prehistoric and even existing animal species to pass back and forth between the continents. Wegener didn't buy either theory.

His own hypothesis was not wholly original. No one looking at an outline of the continents, particularly east and west of the south Atlantic Ocean, could fail to be struck by the correspondences between the African and South American coastlines—the way Brazil's jutting Cabo de São Roque seems to fit, hand in glove, into West Africa's Gulf of Guinea; the way the convex curve of Cabo de São Tomé seems to yearn for the concave bend of the Angolan shore. Such a conclusion is strengthened the more closely one studies the continental edges—particularly, Wegener pointed out, the true continental boundaries: the continental shelves, underwater since the end of the last ice age. He believed that these land masses, with their matching contours, must once have been one. They must have, over millions of years, drifted or been pulled apart—separated "like pieces of a cracked ice floe in water," he wrote—and in places, shoved together, stirred into movement by some force that, admittedly, Wegener couldn't name but which, he was convinced, would someday be identified.

Wegener returned to Greenland in 1912, this time as part of a year-long Danish expedition that would become the first to overwinter on the continent itself. Barely a year after returning to academic life at Marburg, Wegener was sent to the Western Front as a lieutenant in the German Army. Twice wounded and sent home to recuperate, he set himself to articulating his ideas about "drift theory," expanding them from an article to a ninety-four-page monograph that would become his defining life's work.

The Origin of Continents and Oceans was first published in 1915. Its central theory proposed that all of the continents were once—hundreds of millions of years ago—united as a single super-continent. That alone was not a new idea in geology; Austrian geologist Eduard Suess had already proposed in 1861 that a massive continent he called Gondwanaland once covered most of the globe and that parts of it rose and fell over time, creating continents and oceans and leaving behind the marine fossils now so abundant on land. Wegener himself had a very different hypothesis.

Over millennia, Wegener suggested, a single land mass that later came to be known as Pangaea broke apart, the pieces moving away from one another and, in some cases, very slowly crashing into one another. Oceans formed in the gaps between the land masses; mountain ranges formed where continents collided. The process was in fact continuing, he claimed; centimeter by centimeter, the continents were still "drifting."

Like a prosecutor building a case, Wegener provided evidence from a wide range of scientific disciplines. He cited geophysical arguments based on evidence that the Earth may be composed of layers of different materials: a solid outer crust and more viscous or even liquid layers inside. If so, convection currents might animate these warmer, softer layers and spur movement of the crust above it, he wrote. More arguments came from paleontology and biology: evidence of animals and plants occurring together on one continent in certain times through prehistory and then no more. Wegener was particularly intrigued by paleoclimatic evidence: glacial scars carved in rock found at temperate latitudes too warm for glaciers, for instance, and patterns of coal seams—ancient peat bogs—that stop at the edge of one continent and resume directly across the ocean. Many scientists had puzzled over fossil evidence that the location of the magnetic poles had shifted over time—"polar wandering," as the phenomenon was called. Perhaps, Wegener proposed, it wasn't the poles that had moved but the continents themselves, "drifting through the oceanic crust and pushing up mountain ridges ahead of them as they moved."

Wegener's hypothesis "placed an easily comprehensible, tremendously exciting structure of ideas upon a solid, scientific foundation," summarized Hans Cloos, a German geologist and contemporary of Wegener. The theory, Cloos said, "released the continents from the Earth's core and transformed them into icebergs of gneiss on a sea of basalt. It let them float and drift, break apart and converge. Where they broke away, cracks, rifts, trenches remain; where they collided, ranges of folded mountains appear."

Comprehensible—and preposterous, according to most scientists. Criticism centered on the missing mechanism: if the continents drifted, what pushed them? Never mind that the so-called "fixists" had no satisfactory explanation for their own theories; that fact didn't slow Wegener's critics. Even Hans Cloos, an admirer of Wegener and his fellow "mobilists," nonetheless judged Wegener's theory to be wrong.

Wegener was undeterred. He wrote a second edition of *The Origin of Continents and Oceans* in 1920. A third edition, published in 1922, was translated into French, English, Spanish, and Russian, bringing his ideas to a much wider audience and a new universe of critics. Among them were members of the Royal Geographical Society, which convened a conference in London in 1922 and another in 1923 to discuss, and largely dismiss, Wegener's ideas. "A very dangerous idea" was British geophysicist Harold Jeffreys's verdict.

By the fourth edition, two and a half times the size of the first, the tone of the writing was almost breathless as Wegener galloped through his proofs, attempting to cram them all in. But for every convert Wegener won, he seemed to gain two detractors. "If we are to believe in Wegener's hypothesis, we must forget everything which has been learned in the past seventy years and start all over again," wrote the head of the American Association of Petroleum Geologists in 1928—Wegener's point exactly. A German geologist declared that Wegener had failed utterly to prove his theory and advised Wegener "not to honor geology with his presence any longer."

In 1929, the year the robust fourth edition of *The Origin of Continents and Oceans* was published, Wegener was surprised to receive an invitation from the German Research Association to lead an expedition attempting to measure the thickness of the Greenland ice sheet using a seismic sounding technique. It would be the most ambitious Greenland expedition to date. Never had anyone made year-round meteorological measurements in the continent's interior, nor on any large ice sheet anywhere on the planet. Never had a scientific team overwintered in the middle of Greenland. Wegener eagerly accepted, and the following spring he and a large team set sail for the continent.

The expedition was plagued with troubles from the start. A late spring delayed establishment of the base camp by a month, and equipment failures further slowed progress. Mid-September arrived, and the midcontinent meteorological station called Eismitte was still not fully provisioned. With winter closing in, Wegener decided he himself would lead the final 500-mile round trip to finish supplying Eismitte. Driving fifteen fully loaded dogsleds, Wegener, German meteorologist Fritz Loewe, and a team of Inuit Greenlanders headed to Eismitte on September 22—a trek that, under good conditions, should have taken fourteen days.

Thirty-seven excruciating days later, Wegener and two companions driving three empty dogsleds limped into Eismitte. They had endured crippling headwinds, waist-deep snow, a narrowing window of daylight, and temperatures plummeting to −50° and colder, and they had used up or abandoned their entire supply of food and fuel. All of the Greenlanders who had started the trek but one—Rasmus Villumsen—had turned back. Loewe had finished the trek on frostbitten feet; unable to travel further, he would overwinter at Eismitte. Wegener and Villumsen spent two nights at the mid-ice station. Then on November 1—Wegener's fiftieth birthday—the pair rose early, broke out some fruit and chocolate saved for special occasions, packed two sleds with 300 pounds of food and a single can of kerosene, and headed out with seventeen dogs into the foggy twilight.

Villumsen's body was never found. Wegener's was discovered the following April, tucked into two sleeping bag covers and carefully buried in a shallow grave of snow-covered ice 132 miles from Eismitte, about halfway to base camp, the grave marked by Wegener's two skis and ski poles planted upright in the snow. Examination of his body suggested that that he died not of starvation or hypothermia or injury but of a heart attack, possibly brought on by sheer exhaustion. It appeared that Villumsen, finding Wegener dead in his sleeping bag in the morning, had dug a grave and carefully arranged Wegener's body in it, filled the grave with snow, marked the spot, and then resumed the journey alone.

Wegener had died an explorer's death: two thousand miles from home, on a cold continent far across the northern sea. But, in a certain sense, he *was* home. Sixty million years before—if the drift theory he so fervently championed was correct—Greenland and Europe were once part of the same continent. Sixty million years ago, had humans yet appeared, a fellow could have walked from Greenland to Graz.

WEGENER THE SCIENTIST WAS GONE, and with him, any momentum to continue the discussion about continental drift. In fact, debate over drift theory had effectively ended a few years earlier, even before the ambitious fourth and final edition of *The Origin of Continents and Oceans* was published. In the minds of most scientists, the discussion had reached a dead end. Lacking evidence for *how* the continents might have drifted, few scientists were willing to continue hashing over evidence that they *might have* drifted. Still, there was among scientists an undercurrent of growing disenchantment with the dominant earth sciences paradigm. No one was ready to take seriously the notion of drifting continents, but the alternative explanations weren't quite satisfactory either.

In the late 1930s and early '40s, the scientific community's attention and that of the world was hijacked by the start of World War II. Allied naval commanders were still stinging from devastating ship losses to German U-boats back in the early days of World War I, and both the British and the U.S. navies were now eager to learn everything possible about the ocean: what lived there, the chemical composition of seawater, the movement of water, and how things—animals, boats, sound—moved through it. They were also keen on better understanding the ocean basins: their contours and the rock and sediments that composed the seafloor. The U.S. Navy had begun to employ a new technology called sonar—invented toward the end of World War I—to navigate and to track other vessels and was eager to learn more about echo sounding and the window it opened onto the contours of the seafloor. It was also interested in anything involving magnetism in the ocean, which might have applications for submarine warfare. That, plus a better understanding of physical oceanography, might help American mariners find enemy subs and avoid detection themselves. Scientific research in service of these goals became a wartime priority.

Harry Hess was a petroleum geologist at Princeton at the start of the war. For some years he had been investigating gravitational anomalies in the seafloor: areas where the gravitational pull seemed to be weaker than normal, areas—it turned out—such as deep ocean trenches. That work had led him and others to a startling conclusion: portions of the Earth's crust that underlay the oceans had apparently become deformed over time and were continuing to be disturbed by forces they were hoping to identify. He and his colleagues had formulated a theoretical model of crustal plates moving on the Earth's surface, animated perhaps by convection currents under that crust, currents in what might be a softer, more viscous or plastic layer of rock lying beneath the rigid crust and upper mantle. His work had been inspired in part by a physical model, built by a geophysicist at the University of California, consisting of a vat of oil covered by a layer of floating paraffin. Drums physically rotated the oil, mimicking the convection currents found in, for example, a pot of boiling water: the heated water rises up in the middle of the pot and migrates outward and back down along the edges. In the model, the mechanical convection current caused the paraffin on the surface to move outward, to the edges of the pot. Perhaps the Earth worked that way, Hess and others theorized: pieces of the Earth's crust moving, like the paraffin, on a softer or liquid layer below, a layer animated by heat from the Earth's decaying iron core.

And then came Pearl Harbor. Hess promptly put aside his research and enlisted in the Navy, which ultimately put him at the helm of an assault transport ship plying the Pacific Ocean between San Francisco and the Philippines. The USS *Cape Johnson* was equipped with sonar, and Hess was tasked with tracking the vessel's travel routes back and forth across the ocean using the ship's echo sounder, bouncing sound waves off the ocean floor to probe the seafloor's contours.

The detailed ocean floor maps that resulted were of great value to the military. Hess, back at Princeton after the war's end, was interested in them as well. He became one of many individual scientists and marine labs funded by the new U.S. Office of Naval Research, established to promote and fund oceanographic research of all kinds, capitalizing on the big leaps in bathymetry—mapping the topography of the ocean basins—and in scientific instrumentation made during the war. Hess began a close study of the ocean basins, which by the late 1950s had been mapped in even greater detail. He was particularly interested in the Mid-Atlantic Ridge, an underwater range the length of the Atlantic Ocean that was found to have a great trough running through the middle of it. To the surprise of telephone companies planning to lay a transatlantic cable across it, the ridge turned out to be a hotbed of underwater earthquakes as well.

In an internal 1960 report to the ONR, published as a scientific paper in 1962, Hess made a startling proposal. He suggested that mid-oceanic ridges such as that in the mid-Atlantic formed where hot liquid from the Earth's mantle, melted by heat from the Earth's core and pushed by rising convection currents, forced the Earth's crust apart and extruded magma onto the ocean floor, where it spread to either side of the ridge and, as it cooled and hardened, formed new oceanic crust. This cracking open of the seafloor and deposition of new seafloor—"seafloor spreading," as the theory was later called—pushed the Earth's crust laterally away from these volcanically active oceanic ridges at a rate Hess estimated at between one and ten centimeters a year, adding breadth to the seafloor and, consequently, pushing that seafloor outward, toward the continents that surrounded it. That process, Hess suggested, "thickened" the continents: as the leading edges of the seafloor pushed out and bumped up against the continental margins, the seafloor plunged under them, adding more mass to the Earth's continents.

The theory of convection currents in the Earth's mantle, Hess acknowledged, was a "radical hypothesis not widely accepted by geologists and geophysicists." And it wasn't the only radical hypothesis that his theory of seafloor spreading was riding on. Hess had been inspired in part by new data from the emerging field of paleomagnetism, data that also pointed in the direction of continental mobility, but from an entirely different angle.

British scientists P.M.S. Blackett and S. Keith Runcorn had studied minesweeping and demagnetizing of ships during World War II; after the war they began investigating the origins of the Earth's magnetic field, seeking evidence that the Earth's polarity may have reversed—that north had become south, and vice versa—one or more times in the past. French physicist Pierre Curie had already found that magnetic materials in igneous rock—the kind formed when molten magma emerges, cools, and hardens into lava—tend to take on the magnetic field of their surroundings as the rock cools. Blackett and Runcorn set out to collect rocks from all around the United Kingdom. If polarity had reversed, one would expect to find some igneous rocks aligned with today's polarity and other, older rocks with the opposite polarity— and they did. The youngest of their rocks, those formed less than about one million years ago—back when modern humans first appeared, hunting mastodons and running from saber-toothed tigers—all shared the Earth's current polarity. But among rocks older than that, those that emerged as magma during the early Pleistocene era or earlier, they found both normal and reversed polarity. And the rocks' alignment wasn't simply one or the other, north or south. They pointed every which way, depending upon their age and location.

Runcorn then set out to collect and compare rocks of the same age found on different continents: Australia, India, North America, Europe. He found that the polarity of a fifty-million-year-old rock on one continent pointed in an entirely different direction from a rock of the same age on the other side of the world. In fact, unbeknownst to American and European scientists at the time, Japanese geophysicist Motonori Matuyama had already found much the same thing thirty years earlier in rocks he examined in Japan. Runcorn and his colleagues began to theorize that the Earth's polarity may have reversed itself on a regular schedule, perhaps every half-million to million years. (Later research has shown that the Earth's magnetic polarity has reversed many times, most recently about 780,000 years ago, apparently without causing any disruption to life on the planet, and apparently on no regular schedule; reversal has occurred as frequently as twice in fifty thousand years and has held steady as long as forty million years.) The conclusion of Runcorn and his fellow paleomagnetists: not only had the Earth's polarity indeed reversed repeatedly over time, but the orientation of various continents must have shifted around, apparently not all in the same direction, nor at the same time. The continents, in other words, must have moved. The time had come, Runcorn and others asserted, to begin to reexamine the theory of continental drift.

FRED VINE HAD NO PROBLEM with continental drift. He had been keen on the concept even before he entered Cambridge University as an undergraduate in 1958; early in 1962, a few months before starting work toward a PhD in marine geophysics at the same institution, he had heard guest lecturers Harry Hess of Princeton and P.M.S. Blackett from London's Imperial College speak on, respectively, seafloor spreading and paleomagnetic evidence of continental movement, and he was further convinced.

Vine's supervising professor, Drummond Matthews, was conducting research in the Indian Ocean when Vine was set to start his doctoral work, so he launched alone into his first assignment: a review of the scientific literature that had been published about magnetism studies of the seafloor. On his return, Matthews shared his findings with Vine, who analyzed them using a newly developed program for a new research tool—the computer—and incorporated them into his literature survey. The resulting article, "Magnetic Anomalies over Oceanic Ridges," was published under both men's names in the British journal *Nature* in September 1963.

The Vine-Matthews hypothesis, as it came to be known, took Hess's conveyor belt paradigm—continents pushed apart by a seafloor growing in both directions from a mid-ocean ridge—and added a twist: if the paleomagnetists were right about repeated episodes of polarity reversals, then the seafloor ought to act like a tape recorder, "playing back" the history of geomagnetic reversal in bands of older and older rock the farther one looked from the mid-ocean ridges. If magma were spewing out of rifts in the middle of the oceans, causing the oceanic crust to grow outward, and if the Earth's magnetic polarity reverses now and then, given that magma takes on the Earth's current polarity as it cools and hardens, then we should see parallel stripes of alternating polarity in the oceanic crust spreading out from mid-ocean rifts, stripes of newly created crust bearing the polarity of the moment they were created, with the youngest rocks closest to the ridge. Which was, Vine-Matthews asserted, exactly what field geologists and oceanographers had found in oceans all over the globe.

Reception of the Vine-Matthews hypothesis was muted at best, particularly in the United States, where a leading marine geophysicist characterized it as "improbable" and "startling." Even at the Royal Society meeting held in London six months later, in March 1964, to reconsider the theory of continental drift, only one speaker mentioned Vine's and Matthews' paper, suggesting it was inadequate to explain the origin of magnetic seafloor anomalies.

The presentation that generated the most buzz at that March 1964 Royal Society meeting was one by Sir Edward Bullard, another prominent Cambridge geophysicist and a member of the generation that had helped launch a new interest in geophysics—a quantitative approach to the study of the Earth. He had been just eight years old when the first edition of Wegener's *The Origin of Continents and Oceans* was published, just starting college when the first English translation was published. He, too, had focused his wartime efforts on magnetic minesweeping and demagnetizing ships, which had led—after the war's end—to study of geomagnetism.

Like Alfred Wegener—like just about everyone—Bullard was intrigued by the correspondence between the coastlines on either side of the Atlantic Ocean. But using brand-new technology, he was able to take his analysis further than Wegener could have dreamed. Bullard and his colleagues wrote a computer program to compare, with unprecedented precision, the Atlantic coastlines of South America and Africa, and the same for Europe and North America. But rather than comparing the current shorelines, as defined by today's sea level, they examined the continental edges three thousand feet below sea level—what might be considered the true continental margins, where the continental shelf drops off steeply. What they found was a near-perfect fit. The gaps and overlaps they found were no more than fifty-six miles wide—a pittance, given the breadth of the Atlantic Ocean and the continents east and west of it. Bullard's conclusion: a fit this close simply could not occur by chance.

Continental drift was no longer merely one man's opinion. No longer were scientists scoffing at the notion of shifting continents; more and more of them were now asking when, how far, and especially, how? There remained many hold-outs, including Wegener's old adversary, Sir Harold Jeffreys: "The standard actually applied to evidence for continental drift seems to be considerably lower than is usual for a new phenomenon," he sniffed in his published remarks after the 1964 symposium. And plenty of questions remained about drift theory's *sine qua non*: convection currents in the mantle, the presumed mechanism that was causing movement of the Earth's crust. That the crust had moved and was in fact still moving, however, was becoming harder to deny. Summing up the conference, one British geologist noted that no individual fact or piece of geological evidence could prove what was still being called "drift," but—as Wegener himself had insisted more than fifty years earlier—combined, these pieces "can hardly be explained any other way." Forty-one years after the same Royal Society had expressed its "profound regret" that continental drift could not be proven, it gave the theory a cautious nod.

Alfred Wegener's death at age fifty probably had no impact on the transformation of drift theory into plate tectonics; that had required a new generation of scientists to be born, people not bogged down with old ideas, young scientists with better instrumentation and generous government funding. Both of those got big boosts during World War II. And they were further boosted during the Cold War, which set ears to listening all over the world: not just spies tapping phones but seismographs recording every snap, crackle, and pop in the Earth's crust.

It's too bad Wegener didn't live to witness his vindication. He would have been eighty-three at the time of the London conference in 1964. What a sight that would have been: the old polar explorer, perhaps slightly stooped but with the same trim build, perched on a chair in the Lecture Theatre of the Royal Institution, his face creased by years of torment from wind and sun and by a grudging smile, his gray-blue eyes twinkling a silent "I told you so."

5

Mad Rise

CAMBRIDGE, ENGLAND, 1965

I enjoy, and have always enjoyed, disturbing scientists.

—John Tuzo Wilson

FRED VINE WAS ALONE IN HIS OFFICE at Madingley Rise when he heard the door opening below and feet pounding up the wooden stairs—three steps at a time, from the sound of it.

"What are you doing?" demanded the breathless voice ascending the stairs. "Stop!"

At the sound, Vine's long face broke into a wide smile. Wilson was back.

John Tuzo Wilson, a geologist from the University of Toronto, had arrived a couple of months earlier, in January 1965, just a few days ahead of Princeton's Harry Hess, both of them at Cambridge on a half-year's sabbatical. The university's Department of Geodesy and Geophysics was housed in a Victorian-era yellow brick manse known as Madingley Rise, for the knoll on which it sat, some distance west of the center of campus. "Mad Rise," as it was known to its denizens, had been built in the previous century as a residence by a distinguished Cambridge astronomer; since 1955 it had been home to what had become Great Britain's largest geophysics group. Vine shared an office with his supervising professor, Drummond Matthews, in what had been the coachman's quarters, above the stables off the house's rear courtyard, next to the hayloft where the department's sediment cores pulled from the seafloor were now stored. Wilson and Hess were both assigned to the guest office on the ground floor of the main house. By the time Hess arrived, however, Wilson had spread out and made the office, for all intents, his own; Hess, when he wasn't collaborating with colleagues, worked from an office at his sabbatical-year home nearby.

Madingley Rise was a quiet niche in an already quiet campus backwater— quieter still that January, when most of the marine geophysicists were away at sea doing research. Vine, who stayed behind to finish work on his PhD dissertation, had found himself in the enviable position of being nearly alone with two of North America's most esteemed geologists. Better yet, Wilson and Hess were, like Vine himself, "drifters"—adherents to the mid-twentieth century's revival of continental drift theory, and as such, out of step with most of their American and Canadian colleagues. Hess had even gone out of his way, upon his arrival at Madingley Rise, to tell Vine he thought the Vine-Matthews hypothesis was a "fantastic" idea. That was practically a first: unabashed enthusiasm for the theory Vine and Matthews had proposed to faint praise a year and a half earlier. And from no less than Harry Hess, whom Vine had heard speak back in 1962 while he was still an undergraduate and already a quiet believer in continental drift. Even Vine's own colleagues at Cambridge had been chary with their support for Vine-Matthews. Department head Maurice Hill had simply changed the subject when Vine had discussed his theory prior to publication, and Teddy Bullard, while generally encouraging, had quickly declined Vine's suggestion that he might co-author the paper; the hypothesis rested upon too many unproven notions for the likes of Sir Edward. Now, heady with fresh ideas exchanged while bent over the squiggled lines of magnetic profiles spread on a desktop or lingering over sandwiches in the house's pine-paneled coffee room, flattered and emboldened by Wilson's and Hess's interest in his theories about faults and ridges and the movement of continents, Vine had even shared with Wilson some thoughts he had about certain fractures in the seafloor—faults that seemed to be connected to ridges, offsetting them, in the equatorial Atlantic Ocean. Wilson hadn't actually taken much interest in those ideas, or hadn't seemed to—rather than toss them around, or even debunk them, Wilson had, at the time, simply changed the subject.

In fact, Wilson had no sooner settled in at Cambridge before leaving again, trading England's damp, chilly winter for the sunny Mediterranean off southern Turkey, where he'd booked a February sailing trip with his family. Now he was back, and—evinced by his energy racing up the stairs—recharged with more, newer, better ideas. Whatever Vine was doing, Wilson declared as he burst into the office, "It can't be as important as what *I* want to tell you!"

If the Vine-Matthews hypothesis was right, Wilson explained in a rush, and new seafloor was being created at mid-ocean ridges and was pushing out laterally, causing the leading edges of these plates of crust to collide with continental crust, there must also be places where large chunks of crust didn't collide but simply slid alongside one another. It was simple geometry. Mid-ocean ridges, scientists had determined, weren't necessarily continuous along the seafloor. Between these strands of mid-ocean ridge there must be what Wilson now described as "transform faults": faults linking one spreading center to another to create an unbroken, though zig-zag, crust boundary along the seafloor, theoretically on land as well. The Earth's crust at these transform "strike-slip" faults wouldn't move up and down but, rather, horizontally, passing side-by-side. And it would not be a smooth passing, like two well-oiled edges; rather, like any two rough surfaces, these chunks of crust would catch, get hung up for a while, until the strain of inexorable pressure from the spreading center prevailed, causing the caught-up edges to finally give way in a sudden lurch. Sudden, as in earthquake sudden.

Take the seafloor south of Canada's Queen Charlotte Islands, off the American states of Oregon and Washington, for example, Wilson posited to Hess and Vine one morning a couple of weeks after his return from Turkey. The three of them were gathered, as they often were, in Wilson's office, hashing out ideas. Scientists had mapped the long mid-ocean East Pacific Rise, a spreading center, across the floor of the southern Pacific Ocean from Antarctica to where it seemed to dead-end at the coast of southern California. The San Andreas Fault was known to run from the southern California shore, up through California, and almost to the Oregon border. Could the San Andreas Fault be an extension of that line, that boundary? If Wilson was right and the San Andreas was a transform fault between two pieces of the earth's crust, it must connect with yet another spreading center north of it, an as-yet unnoticed mid-ocean rise right off the Pacific Northwest coast, continuing the zig-zag pattern of crust margin along the edge of North America.

"Look, there should be a ridge here," Wilson said, pointing at a sketch map of the known faults along the west coast of North America that was spread on a desktop. Hess, a chain smoker, took a pull on his cigarette and broke in, stabbing at the sketch map and scattering ash.

"Well, that's one of the few places in the world where we have a detailed magnetic survey," he said. "If Fred is right, there should be some expression of the ridge in the magnetic anomalies."

Hess didn't need to be any more specific. Each of them knew exactly which survey and map Hess was referring to.

Thirteen years earlier, in spring 1952, another British geophysicist on sabbatical from Imperial College had had a brainstorm while attending a meeting of the California Institute of Geophysics at a seaside conference center in La Jolla. During a coffee break, Ron Mason had begun chatting with a seismologist from Scripps Institution of Oceanography, wondering aloud about the practicality of towing a magnetometer—a device used to measure the direction and strength of the Earth's magnetic field in specific spots—behind a ship to probe the magnetism of oceanic crust far below. Not that Mason knew anything about oceanography—his expertise was firmly grounded in continental rocks. But magnetometers had been towed behind airplanes during the war, monitoring the continents' magnetism from the sky, Mason observed; what about tracking the magnetism of the seafloor from the sea's surface? His musings had caught the ear of Scripps director Roger Revelle, and three years later, Mason found himself on board the U.S. Coast and Geodetic Survey ship *Pioneer*. The *Pioneer* had been charged with using an echo sounder to make a detailed bathymetric survey of the ocean floor off the West Coast of the United States, from the Mexican border to the southern end of the Queen Charlotte Islands north of Vancouver Island, primarily to improve nuclear submarine deterrence. Mason's magnetometer research could be done at the same time, and his project was tacked onto the vessel's agenda. The ship was to sail back and forth from the foot of the continental shelf to a point 250 to 300 miles to the west—parallel east-west tracks the entire length of California, Oregon, and Washington.

Mason and his assistant, electronic technician Arthur Raff, took part in twelve cruises, the last one in October 1956. They spent the first half of 1957 analyzing the data they'd gathered. The results had been startling—and mystifying. They summarized their results in a map depicting areas of normal and reverse magnetism in, respectively, black and white. They found none of the seeming randomness that Blackett and Runcorn had found in rocks on land a few years earlier. Rather, areas of normal and reverse magnetism on the seafloor appeared as skinny parallel stripes, consistently trending north-south or northeast-southwest the entire length of the map, from Canada to Mexico. Seen in black and white, they resembled nothing so much as the stripes on a zebra. Only in a few places were the long parallel stripes interrupted, mainly at known undersea fault lines. Deepening the mystery, the undersea striping didn't seem to correlate with known topographic features of the ocean floor.

Something important was going on here, some process occurring to create this well-ordered pattern of magnetic anomalies on the seabed. But Mason hadn't a clue what it was.

The paper that Mason and Raff published in 1961—postwar security concerns kept it from public view until then—had been widely read, to little effect. It had remained, in the four years since then, perhaps marine geology's most compelling cryptogram: a pattern of magnetic polarity not the least bit random and obviously significant, if only scientists could interpret what the pattern meant. No scientist in the world had yet found a satisfactory explanation.

Of course, Vine now realized in a rush—*Mason and Raff*. Now it was the lanky Vine's turn to dash up, two stairs at a time, to the library on the house's second floor, where bound volumes of the *Bulletin of the Geological Society of America* were shelved. Vine knew exactly where to find volume 72; he had consulted it many times. Thinking back years later, he couldn't remember whether he actually stopped to take a quick look at the map, by himself upstairs, or had simply grabbed it and dashed back down, laying open the page for his waiting colleagues. The three men stared at the map in amazement, as if seeing it for the first time.

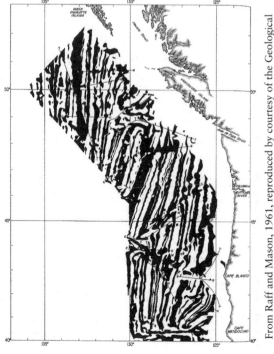

From Raff and Mason, 1961, reproduced by courtesy of the Geological Society of America

Mirror images: that's what Hess meant. Magma oozing up from a hot, molten layer of Earth wouldn't flow out and spill over and harden on just *one* side of such a ridge. It must be creating new crust on both sides of these ridges simultaneously, growing and pushing out laterally to make room for more new crust. The crust that emerged on both sides of the ridge would necessarily be magnetized according to the Earth's polarity at that moment. That meant that the stripes of polarity—be they normal or reversed—on either side of a ridge must be about the same width, reflecting the time span of each particular period of polarity. Thin stripes on both sides for short periods between reversals, fat stripes on both sides for longer periods. If he was right, it should be possible to locate these spreading centers not only by topography—elevated ridges on the seafloor, identifiable by sonar—but by examining the patterns of polarity stripes on that seafloor. If the seafloor was spreading, had been spreading for millions of years, a closer look at the mysterious stripes on the Raff and Mason map might show a pattern of mirror images: same-width stripes—fat and thin—reflecting periods of magnetic polarity of varying length of time—long and short—on either side of a ridge north of the San Andreas Fault and south of the Queen Charlotte Islands. Where that ridge lay should be perhaps the fattest stripe of all: the stripe of our time, the magma on both sides of the ridge reflecting the magnetic polarity of today.

And there it was on the page before them, just as the Vine-Matthews hypothesis had predicted: mirror image squiggles, thick and thin in the same patterns on either side of what was clearly a mid-ocean ridge running at an angle off the Washington and Oregon coasts, and what seemed to be a separate ridge off the southern Oregon coast, the two linked in Wilson's proposed zig-zag manner.

Vine hadn't seen anything like that in the magnetic profiles he'd been scrutinizing in small maps of ocean floor polarity in the north Atlantic and northwest Indian oceans, nor had he expected to. Geology was never that simple. He figured that intrusion and extrusion from basaltic dikes and flows would complicate the polarity at mid-ocean ridges, and it had. But the rate of seafloor spreading in the Pacific, he came to understand, was faster; hence, the blocks of a particular polarity were wider and their margins less contaminated by complicated geology. Here on the Mason and Raff map, a long, thick blob of black ink pointed toward the northern tip of Vancouver

Island, representing all the magma that had spilled to either side of the ridge since the last geomagnetic reversal some 780,000 years ago. Moving outward from it, laterally, were parallel stripes—a narrow band of white, a tracery of black, a thicker band of white, a substantial squiggle of black, another moderate white stripe, and on and on. The map's black-and-white meaning was now clear.

Sonar alone could not reveal the existence of these ridges; sediments coursing down the Columbia River for millennia had filled and smoothed the bathymetry of the ocean floor here. It was the magnetic profiles that told the story, validating Wilson's theory of transform faults while revealing the existence of hitherto unknown mid-ocean spreading centers just west of the coasts of Oregon and Washington, a couple of zigs in the zig-zag eastern boundary of the chunk of Earth's crust underlying most of the Pacific ocean, along two lengths of mid-ocean rise Wilson would soon name the Juan de Fuca and Gorda ridges. Those ridges were the missing links—the last mid-ocean ridges needed to complete the worldwide map of what Wilson would refer to, in a groundbreaking article published in *Science* at the end of that year, as "plates." And one more convincing piece of evidence of seafloor spreading.

Mason and Raff's map of the seafloor off Oregon and Washington had become a kind of Rorschach inkblot test. A thousand scientists had looked at those black squiggles on white paper and seen only zebra stripes. Wilson, with Vine and Hess, saw that and much more: the history of the Pacific Ocean, of the whole Earth, and its future.

6

Live Land

TOM, 1968

IT WAS HOT ON THE SIDEWALK along Hilltop Drive, where Tom Horning sat with the rest of the eighth grade class. Hardly anyone in the neighborhood was home: just one old lady who peered out her picture window at them but didn't offer Tom or his friends anything to drink, not a glass of lemonade, not even water, not even sips from her garden hose. It was cruelly hot, cruel to kids who knew how rare a hot day in May was in Seaside and who could think of a million things they'd rather be doing than sitting cross-legged on the sidewalk along Hilltop Drive with nothing to eat and nothing to drink and nothing to do. They had played baseball for hours that morning, about a million hours it seemed like, before the teachers had rounded up all the middle schoolers and marched them up Broadway, over the bridge across the creek and up to Hilltop Drive with no explanation whatsoever. The teachers were being really tight-lipped about what was going on, but it didn't take a genius to figure that it was probably another tidal wave alert.

Tom and the other eighth graders sat apart from the younger kids, on principle. It was May, so technically they'd be high school students in a month. Teachers played games with the little kids from Central School— "Farmer in the Dell," holding hands and singing—but had left the middle schoolers to their own devices, and they were supremely bored. And hot. And did they mention thirsty? There was no place to pee, either, so maybe it was just as well they had nothing to drink. There were some kids from the high school on the hillside too; they'd driven up in their own cars and in teachers' cars and were mostly on the next couple of streets up, on Hemlock and Alpine—keeping their distance from the middle schoolers, no doubt. Well, that's okay, Tom thought. He knew something about hazing, and just because he was big and tall didn't mean he was spoiling for a fight. Rumors began circulating on the hill, filling the vacuum of information: a tidal wave had already struck the town, or teenagers had escaped down the hill to surf the incoming wave. Tom didn't doubt it. He hoped it was a tidal wave. No tidal waves in the whole history of Oregon, then two before he even hit high school? Things in Seaside were definitely looking up.

It was past 3 p.m. when the teachers finally asked everyone to line up again, and the whole crowd on Hilltop Drive—almost the entire Seaside population ages five through fourteen—marched back down the hill to Central School and Broadway Middle School, where buses were waiting to take them home. Just another false alarm, people said. There had been an earthquake, all right; Tom, already a newspaper enthusiast, would read about it the next day. A big one in the ocean east of northern Japan, the Tokachi-Oki quake, one of the worst in recent Japanese history; it had killed dozens of people on the islands of Honshu and Hokkaido. But concerns about a cross-Pacific tsunami hadn't panned out, the newspaper reported. Only a minor wave had been registered on the shores of Adak Island in the Aleutians, northeast of the quake's epicenter; ditto for Guam, southeast of Japan. It was a disappointment for the crowds that had flocked to the oceanfront Promenade in Seaside and to the beachfront dunes west of Gearhart and the hundreds of motorists clogging U.S. 101 seeking shoreline vantage points. More cautious residents, including carloads of high school students, had fled to Crown Camp or at least to the taller sand ridge in Gearhart. But by late afternoon, the staff of Seaside Hospital had begun moving patients back down to their rooms on the ground floor, which had been evacuated. Coast Guard helicopters that had been moved upriver to Knappa began returning to their base in Astoria. And police and civil defense authorities began fielding the usual barrage of calls complaining about yet another false alarm.

False alarm? Tom wasn't so sure. As soon as he got home from school that day, he headed across the yard and down to the bay, where he went every afternoon when it wasn't pouring rain. Tom's engineering prowess had grown considerably more sophisticated in the four years since he was gouging truck caves out of the riverbank. Now he was constructing levees, dikes, dams across portions of the bay below his house, single-handedly challenging the moon's gravity and the momentum of the ocean. By spring of eighth grade his hydrology efforts had progressed from a few sticks and stones to placing entire drift logs across the little channel running between the shore and a small sandbar some twenty feet away. Sweating and straining and kneeling in the sand, working alone, he'd start by wrestling logs into position perpendicular to the channel, sliding them on the slick, dark bay mud, then pounding in sticks to stake the logs into place and packing the works with mud and sand, working like a beaver. He had only a small window of time in which to work: one of the twice-daily low tides, whichever one fell when he wasn't asleep or in school. Typically he worked after school, hurrying sometimes to throw a few more logs on the dam before the tide turned, working as the water rose to his thighs, to his waist, flailing his arms like a dredge, sweeping up mud and sticks and stuffing them in cracks to stanch the flow. As satisfying as the construction itself was, watching the tide destroy his work was just as exciting: the initial creeping of water, the inexorable rise, until finally the river was fully running backward, toward the mountains, as if the Earth itself had tipped and the ocean was pouring in. Watching the dam hold was good. Watching it collapse was even better for a fourteen-year-old boy attuned to disaster, picturing the imagined death and destruction following the failure of his mini-Hoover Dam. He'd get back to the house reeking of septic tanks and black slime and be forced by Grandma Lear, the housekeeper who was in charge while Bobbie was at work, to hose off on the back porch, peeling off his stained jeans and dumping them directly into the washer. His white socks and white Converse All-Star tennis shoes had taken on a gray cast that spring that the most vigorous scrubbing couldn't touch.

As soon as he got to the bay that afternoon, Tom could see right away that something unusual had happened while he had been killing time on Hilltop Drive. The tide was out but was on its way back in, so the sandbar should still have been white with dry sand on that hot day. But it was gray, freshly wet, as if a wave had washed over it. And there was something else. The bottom of the channel where he built his driftwood dams was covered with two or three inches of pale beach sand, obscuring the foot or more of black mud that normally blanketed the bottom of the channel, the mud responsible for staining his socks. He'd seen sand like this, out-of-context ocean sand, once before, four years earlier, when the tidal wave from Alaska had turned his own grassy yard into a beach. Now the tide had just begun to creep into the channel, and he found that if he stood on the fresh layer of sand at the edge of the channel, above the waterline, it flexed like a big, thick sheet of rubber, firm on the outside and squishy on the inside. Tom ran to fetch his little brother.

"Walk on this!" he said, pointing David toward the fresh sand, and the two of them stood and stomped on what Tom started calling "live land," laughing as it slid this way and that under their feet. They found that the sand would support their weight for a few steps, but if they ground their feet into it, it would flex and lose strength; water would seep up through the sand, causing it to collapse in on itself.

Tom could see that a tidal wave, or something like it, *had* entered the bay mouth and swept up Neawanna Creek, no doubt about it, whatever anybody else said, even the newspaper. Years later, he would learn that he hadn't been the only witness. Logger-turned-taxi driver Ab Burke had been parked at the viewpoint across the street from the high school that afternoon and had watched as a low wave, two or three feet tall, entered the bay and marched like a tidal bore up the Necanicum River.

But that afternoon, in the waning days of eighth grade, Tom kept his observations to himself. He felt no need to announce to the world what he knew. He was the fourth of five children and had learned to keep his own counsel. Lots of cool things happened out in the bay, every day. No one else seemed particularly interested in any of it.

7

Pay Dirt

An earthquake.
A small flex of mobility in a planetary shell so mobile that
nothing on it resembles itself as it was some years before, when
nothing on it resembled itself as it was some years before that,
when nothing on it ...
— John McPhee, *Assembling California*

TANYA ATWATER STOOD AT THE FRONT of the conference room at Asilomar, listening as the moderator recited her pedigree: Massachusetts Institute of Technology, University of California at Berkeley, University of Chile Geophysics Institute, Scripps Institution of Oceanography. As she waited, she scanned the crowd through wire-rimmed glasses: a sea of white faces, some bearded and some clean-shaven, sweaters and buttoned shirts and polos in white and tan and a muted palette of mostly blues and browns. She herself was wearing the ankle-length red velvet dress she'd bought on a recent trip to Mexico City for another scientific conference. No love beads around her neck; they didn't work with the dress. For this occasion, she'd opted to wear shoes—it was December, after all.

The audience of not quite one hundred scientists was much smaller than the one she'd addressed at the American Geophysical Union conference. For that presentation—her first real professional talk—she had been thoroughly coached, had rehearsed over and over. Now Atwater, in her fourth year of graduate school at Scripps, found herself wishing she'd practiced this talk a couple of times; she'd been too busy working and reworking the paper behind it, one in which she attempted to explain the origins of the San Andreas Fault and, in that process, untangle the tectonic evolution of all of western North America over the past sixty million years or so. Too bad her presentation was scheduled for the fourth of the conference's five evenings; not until it was over would she really relax and join in the discussions that had been spilling out of the scheduled conference sessions and onto the landscaped grounds and nearby beach and into participants' rooms, discussions sometimes raging far into the night. There was nothing twenty-seven-year-old Atwater liked more, from the moment she opened her eyes in the morning to when she closed them at night, than to talk science with somebody.

Most of the participants at this conference—the first Penrose Conference of the Geological Society of America—were established scientists midway or further into their careers. Only a handful of graduate students had been invited to take part, Atwater among them. "The Meaning of the New Global Tectonics for Magmatism, Sedimentation, and Metamorphism in Orogenic Belts," organizer William Dickinson of Stanford University had titled the meeting, shamelessly borrowing wording from the title of a seminal paper—"Seismology and the New Global Tectonics"—that a trio of seismologists from Columbia University's Lamont Geological Observatory had published the year before. The idea of the Penrose Conference, named in honor of an early GSA benefactor, was to provide an intimate forum—in a relaxed setting, and on a relaxed schedule, stretching over several days—for an informal exchange of information and ideas in geology and related fields, stimulating discussion and collaboration in a way that's difficult to do in larger scientific meetings open to all comers.

It was the *intimate* part that had most stymied Dickinson. His intention had been to limit participation to just fifty scientists, certainly no more than seventy-five. In the end, Dickinson had issued invitations to some ninety-five hand-picked participants from around the world, put them up at the conference center on California's Monterey Peninsula for six days, and invited them to share their findings, evangelize, and listen, discuss, and debate.

The crowd waiting to hear Atwater speak was a friendly one—friendly, and impressive. Atwater was one of five grad students and one faculty member who had made the trip up from Scripps. She could make out the faces of the Stanford University bunch, Dickinson himself and a couple of grad students: she knew each of them and their work, and they hers. There was one geologist from MIT, where she'd started college, and one from UC Berkeley, where she'd completed her bachelor's degree after deciding she really needed to be studying geology out west, where the rocks were young and fresh, practically still warm. George Plafker from the USGS was there, chatting to her about geologizing in South America.

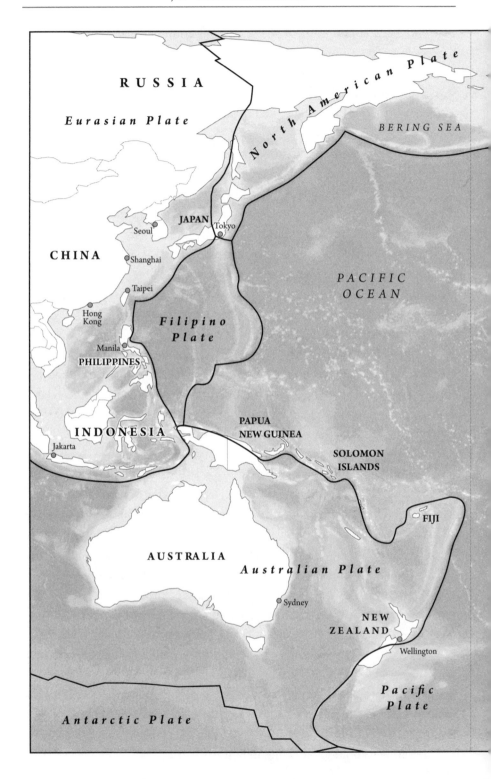

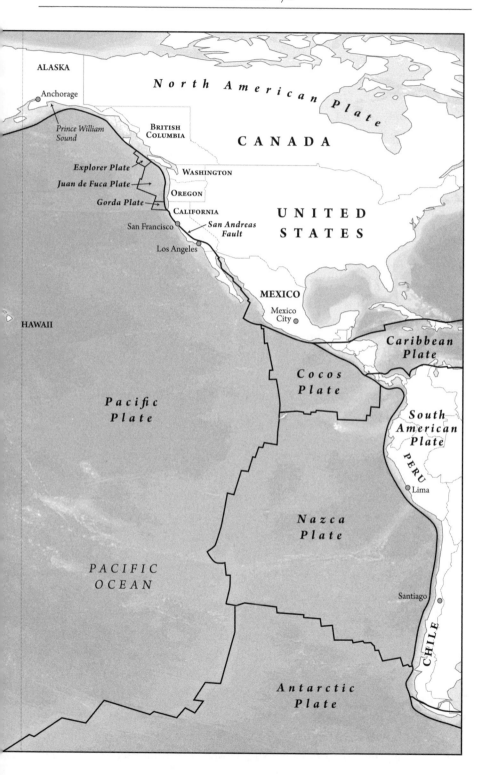

Lynn Sykes was there too. Atwater and her fellow grad students had devoured his "Seismology and the New Global Tectonics" paper, had debated it for days, and were still talking about it. Sykes's work—correlating earthquake activity with developing models of plate movements—was a happy outcome of the Cold War nuclear anxiety that had followed World War II. The U.S. Coast and Geodetic Survey had, in the 1960s, developed what it called the Worldwide Standard Seismic Network in an effort to detect nuclear bomb tests wherever in the world they might be set off. But the new, improved array of uniform seismographs deployed as part of the WWSSN didn't just pick up bomb detonations; they were equally adept at recording the Earth's own bumping and grinding and, in so doing, supplying seismologists with an unprecedented level of detail about the location, depth, magnitude, and movements of earthquakes all around the globe—earthquakes that, it was now clear, were by-and-large associated with plate boundaries.

John Dewey from Cambridge University and his research partner John Bird from SUNY were there, and Ken Hsu of Zurich Geologisches Institut, and Tohikiko Matsuda and Seiya Uyeda from Tokyo University's Earthquake Institute: many of the brightest minds in the business. All men, all but Atwater herself. But that aspect of geology was something she hardly noticed any more.

Atwater had been in South America in 1966, taking a break after Berkeley for some overseas adventure and working as a technician in the Geophysics Institute at the University of Chile, when she realized a revolution in the earth sciences was under way and she was missing it. Jim Heirtzler of Lamont Geological Observatory had happened to be passing through Santiago that summer and stopped to give a talk about the latest theories of seafloor spreading based on brand-new geomagnetic data from the ocean floor at the Pacific-Antarctic Ridge. Heirtzler displayed what was labeled the *Eltanin*-19 magnetic anomaly profile: a horizontal line of sharp up-and-down zig-zags representing areas of normal and reversed magnetism on the seafloor on either side of the Pacific-Antarctic Ridge between Antarctica and New Zealand—a hypothetical spreading center—recorded during the nineteenth cruise of the research vessel *Eltanin*. Gazing at it, Atwater felt her hair stand on end. There it was, solid evidence for seafloor spreading: undeniable, as far as Atwater was concerned. Atwater had seen magnetic profiles before, including the Raff and Mason map published in 1961, but

Eltanin-19, with its mirror-image peaks and valleys rising to a broad, high crest at the center, site of the ridge itself, was the clearest, most symmetrical profile she had ever set eyes upon. Immediately she began making plans to leave Chile and return to the United States for graduate school, desperately hoping that she hadn't missed all the fun.

In fact, her timing was excellent. She arrived at Scripps in January 1967 and learned that Fred Vine had visited just a month earlier and given a talk attended by the entire faculty and student body. Vine had shared the *Eltanin*-19 and other magnetic anomaly profiles and, as Atwater heard it, converted the last remaining faculty fixists to mobilists in the course of his talk. By that spring, Atwater was on a boat off the northern California coast, recording magnetic polarities at the Gorda rift.

By summer 1969, Atwater's attention had both broadened and narrowed to focus on the origins of California's San Andreas Fault, which she and others now believed was a significant boundary in the global system of tectonic plates. Then began what would become the most intense period of work in her life, her first major solo writing effort, which led to her talk at Asilomar and, ultimately, to publication as the lead article in the December 1970 volume of the *Bulletin of the Geological Society of America* and to a chunk of her PhD dissertation. She rarely ate, she barely slept all that summer and fall. The work was all-consuming, not only because the writing was itself time-consuming but because of the sheer complexity of the concepts, and the newness of the data and the assumptions—of seafloor spreading and plate tectonics—upon which it was built. She was tackling the entire tectonic history of western North America: the way the plates there had converged and diverged, pulled apart from one another and scraped by one another and consumed one another, forming rifts and ranges over tens of millions of years, shaping and reshaping the continent. High drama, played out over a stretched-out geologic time scale: magma bubbling out of mid-ocean ridges, enlarging plates that migrate toward continents where the plates dive under older, more buoyant pieces of crust, falling farther and deeper into the Earth until they melt the mantle, which bubbles back up as magma to feed volcanoes that erupt and create new landforms, the pattern repeating and shifting slightly, and the look of the land changing with it.

The Asilomar conference had come at the perfect time for Atwater. After months of writing, after vetting her paper with mentors and colleagues and incorporating their feedback, she was ready, by December 1969, to take it to a bigger stage before submitting it for publication.

The concept of subduction had been the key, or one key, to putting the whole North American story together: the process by which one plate thrusts under another, an oceanic crust diving under a continental crust, the rocks of that descending crust diving deep, halfway to the hot center of the Earth. But no one had called it subduction until a couple of days before Atwater's own talk at Asilomar. "Mantle underflow," her advisor at Scripps had called the process. *Awkward*, pronounced research scientist Dietrich Roeder in an evening bull session fueled equally by passion and Jack Daniels. Furthermore, Roeder declared, what would you then call the place where this "mantle underflow" took place—the *zone of mantle underflow*? No, no, no—surely, Roeder insisted in his sibilant Swiss accent, English-speaking geologists could come up with a more eloquent term for this process and for the places where it occurs, places like the ocean floor just off Chile, where the oceanic Nazca Plate converges with the South American Plate to create the Andes Mountains, or the various plate convergences off Japan. Happily, Roeder exclaimed, he and his colleagues at Esso Production Research Company in Houston were prepared to propose alternatives. They had, in fact, come across a couple of likely candidates in a search of the literature of Alpine geology.

"*Verschluckungzone!*" Roeder had declaimed—literally, *swallowing-up zone*—offering a term coined shortly after the turn of the century by an Austrian geologist, a countryman and contemporary of Alfred Wegener himself. That would never fly, Roeder could see from the looks of mock horror on the faces of his assembled colleagues. So Roeder offered up another, likelier candidate, one used by a French geologist in the early 1950s: "*zone de subduccion.*" Roeder's own recommendation, greeted with cheers, was to Anglicize the French term and call it a *subduction zone*. By week's end, the term had been codified by the community in the lightning-fast pace of change that characterized geology in those days: blink, and you'd find the ground itself or your perception of it or the name you used for it—or all three—had shifted.

The concept, if not yet the term, was embedded in Atwater's opening remarks, which she launched into just after 7:30 p.m. that Thursday evening. The San Andreas Fault as we know it today is a relatively new phenomenon, no more than thirty million years old, she proposed. It is a "strike-slip" fault: a place where the North American and Pacific plates are moving past one another in fits and starts, expressed in frequent earthquakes. A study of magnetic anomaly patterns in the northeast Pacific Ocean, when viewed through the lens of plate tectonics theory, suggests that in the vicinity of today's San Andreas Fault there was once a subduction zone—she would call it the Franciscan Subduction Zone—that evolved into a different kind of margin as tectonic plates, some long since subducted under the North American continent, moved and grew and disappeared. To understand the complexities of continental geology, she suggested, you need to look to the ocean floor, to the ever-growing oceanic plates. And to understand the ocean floor, it's essential that you look at the continents, at what that oceanic crust has wrought.

Take the Juan de Fuca and Gorda ridges off Washington and Oregon, for example, she said. The combination of seafloor magnetic anomalies offshore and the volcanic Cascade Range a hundred miles inland suggests that the ocean floor between the continent and those offset, parallel, underwater ridges is acting as a single plate—call it the Juan de Fuca Plate—thrusting under the North American Plate: *subducting* under it, if you will. Today the Juan de Fuca Plate is tiny, one of the smallest plates in the worldwide system of plates, she said, but it was not always so. The Juan de Fuca Plate (including neighboring chunks of the Earth's crust that would come to be called the Gorda and Explorer plates) is almost certainly the last remnant of a huge plate, the Farallon Plate, at one time larger even than the massive Pacific Plate but which, over time, has been almost entirely consumed by millions of years of subduction under and into North America.

"Okay, thank you, Miss Atwater ..." the moderator eventually broke in. Atwater, panicked and a little chagrined, checked the clock on the wall; she wasn't nearly finished with her presentation and was already way over time. But before she could open her mouth in apology or plea, a voice boomed out from the audience. "Aw, let her go on! This is great stuff!"

In fact, the crowd was rapt—"blown away" was the term one British geologist would use years later to describe his and others' reactions to Atwater's session at that pivotal conference. So the moderator stepped aside, and Atwater did go on in her detailed, precise manner. And when, nearing the end of her talk, someone spoke up to challenge her assertion about the young age of the San Andreas Fault, another voice rang out.

"It's true! It's true! Believe it!" It was Ken Hsu, the distinguished Chinese-born, American-educated, and now Swiss-based geologist, jumping to his feet. Hsu had been among the scientists aboard the research vessel *Glomar Challenger* when it transited the South Atlantic Ocean the previous year between Dakar, Senegal, and Rio de Janeiro, Brazil, on the pivotal third leg of its epic eighteen-month cruise for the Deep Sea Drilling Project. The project, managed by Scripps, had set out to test the hypothesis of seafloor spreading and plate movement by age-dating the fossils found at the bottom of piles of sediments at different sites on the ocean floor. The deepest sediments, those that lay on top of the "basement rock" created at a mid-ocean spreading center, should be about the same age as the rock itself. By dating the deepest fossils found at varying distances from the Mid-Atlantic Ridge, and comparing them to the ages geophysicists had projected for that same site based on its magnetic polarity, the project amounted to an affirmation of the theory of seafloor spreading, a test that, as Hsu put it, hit "scientific pay dirt."

As Hsu spoke, gushing about his own conversion to mobilism aboard the *Glomar Challenger*, Atwater took a moment to scan again the sea of faces, many of them men twice her age—legendary figures from leading research institutions—and others, like herself, just starting their careers. It was a good time to be a young geologist, maybe the best time ever, she mused. It was especially good to be an oceanographer or marine geologist, exploring the largely unknown ocean floor with a fresh and flexible mind unburdened by the baggage of the old pre-plate tectonics geology.

She had been wrong in Chile; that she now knew. She'd been in such a hurry to get back to the States, so worried that all the important work had already been done, the key discoveries already been made. In fact, she'd arrived at Scripps at the perfect time, with the right tools and the right turn of mind to add her own pieces to the puzzle and, in that process, help redefine the field and set its new direction.

Atwater had not missed the revolution, as it turned out. She was, happily, right in the thick of it.

8

Our Turbulent Earth

TOM, 1972

BY SPRING 1972 THE TSUNAMI FROM ALASKA was a distant memory for most people in Seaside. A new single-family house had been built on the site of the wrecked Finnish Meeting Hall across the street from Tom's house. The Jacksons' house had been torn down as well, replaced with waterfront townhouses.

Seaside's city leaders were bent on looking forward, not back. A big new convention center had just opened downtown between the beach and the river, positioning the city to welcome not only individual tourists in summer and during spring break but conferences year-round. A new hospital had been built on the east side of town, across Neawanna Creek—a much larger, more modern facility, at a site less likely to flood in a tsunami or winter storm. The Seaside School District had moved its administrative offices into the old hospital building on the banks of the Necanicum River downtown and was moving ahead with plans to build a new elementary school on the hillside near the new hospital, replacing Central School downtown. A traffic light—the city's third—had been installed where Broadway crossed busy U.S. Highway 101, and there was talk of building a municipal swimming pool.

There had been big changes at the Horning household too—now the Horning-Jensen household. In 1967, Bobbie Horning had married the family's bachelor neighbor, Jim Jensen, the same neighbor who had been out poking around in his pajamas with the Horning kids the night of the 1964 tsunami. The couple's timing could have been better; with four teenagers and an eight-year-old at home, never had the house on Twenty-Sixth Avenue felt so small.

Jensen was wise enough to leave discipline and other parenting decisions to Bobbie. Jensen wasn't really the fatherly type; even Tom, thirteen years old, could see that his new stepfather was at a bit of a loss in his new role and a little overwhelmed by the sheer press of humanity in the house. It was his first marriage; he had no kids of his own. He did the things he seemed to think a father should do, taking the boys fishing on the Columbia and car camping on the southern Oregon coast, even taking a jetboat excursion up the Rogue River, where they tried panning for gold. Jensen considered himself a desert rat at heart; he had spent much of his life rambling in the desert of southern California and had moved to Seaside only to keep his widowed mother company. He barely tolerated the North Coast climate, with its long, wet winters and breezy summers, its fleeting blue skies, and he often took off by himself to Oregon's High Desert, where he'd spend days or weeks prospecting and camping alone.

Jensen wasn't as well educated as Bobbie, a difference Bobbie seemed willing to overlook but one Tom noticed. He was quiet and didn't try to insinuate himself unnecessarily into the Horning kids' rhythm. But there were little things. His deafness, for one: he wouldn't acknowledge it and wouldn't do anything about it, and he'd get embarrassed and angry when he realized he'd missed part of a conversation, sometimes hinting that he'd been intentionally excluded. But for the most part, he rolled with the punches, biding his time until all the kids left home.

Which they had already begun to do. Lynne started college that year and never again lived at home. A year later, Judy graduated and left, eventually becoming a nurse like her mother. About that time Jensen's own mother moved back to California as well, leaving empty the little house Jensen still owned next door. It became a workshop, a storage space, and eventually a crash pad for returning Horning kids.

Soon only David would be left at home. Tom had been accepted at Oregon State University in Corvallis, two and a half hours away by car. He declared chemistry his major, inspired in part by the wonder stirred in him by Grandpa Baker's gift of the silver dollar almost a decade earlier, a gift that had prompted the question he continued to roll around in his mind: *where does it all come from?* Plenty of college departments could take a stab at that question, but he liked the way chemistry approached it—quantifiably, tangibly, and a little dangerously. He'd enjoyed high school chemistry lab, led by an inspired teacher, but many of the chemical solutions they experimented with came pre-mixed, and Tom found himself wondering what their story was. Where does strontium nitrate come from? What about calcium chloride; how do you get it?

Popular Science magazine was one place where Tom's interests overlapped his stepfather's. Jim Jensen was a sheet metal worker by trade, a handyman and a tinkerer who subscribed to the magazine to keep up with the latest thinking about how things worked. Tom looked forward to the magazine's arrival and would read every word. The May 1971 issue had an article about how to make your own slide rule, and another about a ship called the *Glomar Challenger* and the Deep Sea Drilling Project: way cooler than the failed Mohole project of the 1950s that they'd talked about in school. According to the article, findings from the Deep Sea Drilling Project seemed to confirm the theory of continental drift—a theory that had always made sense to Tom, just looking at a map of the world.

But it was the May 1972 issue—the month before Tom's high school graduation—that really caught his attention, taking the *Glomar Challenger* story to its logical conclusion with language that spoke directly to a curious young man with a boy's sense of wonder. "Throw away the geology textbooks," read the subhead over the magazine's lead story. "The surface of our 'stable' planet is in constant motion, with the very continents whirling, tilting, drifting, and plunging beneath our feet." The title began with a two-word phrase Tom had never heard before.

"Plate Tectonics," it read: "A Startling View of our Turbulent Earth."

9

Wave of the Future

TOM, 1976

THE DISPLAY CASE OF SPECIMENS in the main corridor of Wilkinson Hall, on the Oregon State campus, seemed to Tom to have an almost magnetic pull, its own field of gravity. Never did he walk by it, on his way to or from class, without stopping to take it in, or at least swiveling his head for a glance. Big quartz crystals, some rose-colored, some purple. Water-clear and milky-white calcite crystals, like chunks of faceted ice. Spiky fingers of gypsum crystals and nests of bronzy nickel sulfide hairs on a bed of white quartz. Mirror-like epidote, its usual pistachio green color darkened, in this crystal, to the green of cooked spinach by what Tom knew had been an infusion of iron leached through rock sometime in the distant past. Actually, to Tom's eyes—he was red-green color blind—the epidote in this crystal was nearly black, a muddy black. But color wasn't the only thing to love in these minerals: there was the texture, the cleavage, the shape of the individual crystals. Window shopping, it felt like to Tom: the pleasure of just looking, unburdened by desire.

This spring day he stopped to take a closer look, drawn by a fist-sized cluster of shiny, boxy pyrite crystals almost golden in color, larger and shinier but otherwise just like the chunks of pyrite he had found as an eleven- or twelve-year-old kid, picking through the rocks at the top of the quarry above the dam on the Lewis and Clark River, south of Crown Camp. A friend of his mother's had told her about the quartz and pyrite you could find around the concrete dam, and Bobbie had driven Tom and his siblings up there, more than once, to clamber up the blasted cliff face above the dam and scramble in the rock piles at its base, digging for rock crystals not unlike the salt and sugar crystals Tom liked to grow in jars on the kitchen counter. Here in the polished glass of the Wilkinson Hall display case, Tom could see his own reflection as well, merging with the minerals: photo-gray aviator-style glasses perched on a strong nose, longish brown hair brushing his collar, his mouth relaxed into that look of wonder leftover from childhood, lips parted slightly in a distracted smile.

The case was well lit and clean, with museum-caliber minerals artfully arranged. Specimens were regularly swapped out, like sculptures in a gallery, by the university's resident volcanologist to create ever-changing, thoughtfully curated displays. Nothing like the minerals in the case that hung upstairs in Gilbert Hall, above the chemistry building's main lecture room, a five-minute walk across campus and, to Tom, a universe away.

Never had Tom seen any indication that the Gilbert Hall case had ever been opened. Glass windows and white painted shelves were held together in a tired wooden frame, and on those shelves sat lumps of naturally occurring minerals, slowly decomposing and collapsing onto themselves. Selenite gypsum crumbling into a pile of silky white straw. A drab sample of charcoal-gray magnetite, its luster masked by a film of dust. A piece of realgar—an arsenic ore—on its last legs. Realgar, Tom had since learned in mineralogy classes, was vulnerable to daylight; the chunk in the Gilbert Hall display case had long ago decrepitated from cherry red crystals to an orangey powder, collapsed into a sad little pile of neon-colored dust.

The chemistry major thing had not worked out; Tom had known it by the end of the first week of classes. Cavernous laboratories reeking of acetone, the smell so pervasive it clung to his clothes and imprinted on his nasal membranes. Pale-skinned students milling about in white lab coats—it just wasn't his crowd, his scene. He'd stuck with the curriculum through freshman year, but by spring he was desperate for a change.

Nuclear engineering had been his mother's idea. "It's the wave of the future," Bobbie had said. Tom liked the sound of it: exotic, cutting-edge, a little prestigious. Tom told his roommate what he was thinking about, and both of them wound up switching majors. But once classes started sophomore year, Tom began to sense he'd taken another wrong turn. His roommate had signed on to the more practical engineering track—coolant pumps and flow rates, nuts and bolts. Tom, plagued as ever by the big questions, had chosen the more theoretical side of nuclear energy, which turned out to mean non-stop calculus classes. His roommate was having fun. Tom was not. He felt like he was always playing catch-up, taking too many challenging courses at once—radiochemistry, upper-division engineering and physics—in an attempt to mesh with his new major's course sequencing so he could graduate sometime in the foreseeable future. Between that and a high-maintenance girlfriend and beer-drinking obligations on the weekend,

Tom was losing balance. Academically, he was getting by. But by spring of his junior year, Tom was burned out.

After a summer of painting houses in Seaside, Tom returned to campus for what should have been his senior year, but he withdrew after a few weeks. His heart just wasn't in it. He spent the rest of that fall bouncing between Corvallis and Seaside, picking up odd jobs and thinking back to a geology class he'd taken on a whim the previous year. The professor, a geochemist, had been a crack-up: funny, irreverent, and smart, peppering his lectures with jokes borrowed from late-night television that kept the class engaged while getting his point across. The whole atmosphere was so different from nuclear engineering, so much more down-to-earth, so comfortable. Besides, Tom thought, what was one of the biggest concerns the public had about nuclear energy? Nuclear waste. How can you tell if the waste storage is safe? Look at the geology. So he went back winter term and, in a bold stroke, registered for the introductory geology curriculum.

It was the mineralogy class that hooked him first—minerals like those in the Wilkinson Hall display case, so unlike the neglected minerals mouldering away in Gilbert Hall. Sparkly rocks, golden and silvery rocks, shiny and coarse, striated and speckled, rocks composed of minerals whose genesis stretched back over hundreds of millions of eventful years, minerals with stories the rocks seemed to be bursting to tell. He could almost feel the silver dollar in his hand again, smooth and cold, his key to the universe.

He was sold. Switching his major to geology meant practically starting over; it would take Tom another three and a half years to graduate. But he didn't care—it just meant he'd have to spend a few more summers at the beach painting houses. Anyway, he'd already completed calculus and most of the other less-fun requirements. All that was left to complete his degree in his new major was geology and more geology.

It was like eating nothing but candy, he mused as he strode down the main corridor of Wilkinson Hall, heading out the door and into the Willamette Valley spring afternoon. *Candy every day.*

10

Epidote Dreams

Tom, 1979

Tom's second geology job after college was almost perfect. It was a summer job in Alaska for a big mining company whose mining claims, scattered throughout the state's interior, had to be periodically "proved up" in order to maintain them. Every morning a helicopter would arrive to pick up Tom wherever he was staying—usually a fancy hunting lodge deep in the outback, like something from a movie. Tom and another geologist or two would climb into the helicopter, and off they'd fly to a claim block somewhere in the Alaska or Brooks Range. There he would spend the long summer day unraveling the geology from one site to the next, digging test pits or drilling holes or sampling rocks and drawing maps, looking for signs of ore that his employer could turn into money, walking many miles in the process. It was real geology, the kind he'd dreamed about: hiking around the mountains of Alaska, looking at rocks, and getting paid to do it.

The malachite was the only problem. Actually Tom's color blindness was the problem. Malachite is a mineral the green hue of bread mold or a well-watered lawn. Malachite is also the surface clue to copper underground. Copper wasn't the only mineral Tom was supposed to be looking for, but it was an important one, given the high copper prices of the day. His color blindness hadn't seemed like a big problem in college. There, he could usually identify minerals by a process of elimination, or by focusing on a rock's other qualities—its cleavage, its shape, its hardness, its habit, whether boxy or lenticular or stellated, depending on the environment where it formed. But now Tom was earning a paycheck based, in part, on his ability to notice the green rocks among the gray. Never before had he understood what a liability his color blindness would be in his chosen career as an exploration geologist, looking for sometimes subtle clues in the colors and configurations of the rocks and the lay of the land, clues pointing to where someone might be able to make a lot of money.

It had become clear to Tom one day late that summer as he was hiking across a mountain slope east of Kotzebue with one of his colleagues. Under their feet—not too far underground, they both knew—lay a seam of pure copper ore twenty-five to fifty feet thick, worth about $5 billion. Tom's company was already working on the economics of extracting it. "*There's some copper,*" the other guy said, pointing at the ground as he and Tom worked their way across the mountainside. "Where?" Tom asked. The other geologist looked at Tom, baffled. "Right there," he said, stabbing a finger. Tom felt like a fool. It all looked the same color of gray, though one part looked as if some egg white had spilled on it. *That* was the malachite, Tom realized; he could walk right over a multibillion-dollar ore body and never even see the clues. He was even screwing up the geology maps he was drawing, color-coding certain geologic units purple when he should have used light blue; it was all the same to him until someone pointed out his mistakes. Never had he felt so incompetent.

Never, that is, except on his very first geology job, the one right before this one, his first assignment with the same mining company. He had graduated from Oregon State University in March and immediately landed a job prospecting near Pioche, Nevada, near the border with Utah. Pioche, a historical mining town with "Old West charm" unsullied by commercialism, according to the chamber of commerce: *Right,* Tom thought, *because nobody in their right mind would vacation here.* He hardly spent any time in Pioche. He ate dinner and slept there, but every morning he and the crew would drive forty miles out of town to look for gold, or rather, for rocks that suggested there was gold nearby. Tom knew the how the geology worked: ore bodies are formed, usually, by hot water in volcanic rock, and that combination produced a variety of minerals: copper and molybdenum in the center of the system, lead and zinc in the halo, and gold and silver in veins out on the edges. His job was to seek the outer fringes of that system, where hydrothermal alteration had transformed one kind of rock into another, rearranging the minerals inside. One clue to finding gold was to first find epidote, a pistachio-green mineral that appears in and on the rock. But Tom's eyes couldn't distinguish the green epidote from the gray rock it sat in. Up close he could identify the mineral by its shape, its appearance in the rock. But not from ten yards away, out the window of a truck going thirty miles per hour.

"Look at the beautiful epidote in that rock!" his boss would exclaim, baiting Tom, who would strain to spot it. Tom had quickly become the laughing stock of the outfit. His boss—who must have flunked out of the School for Effective Management, or never enrolled—brought him weird rocks to identify, minerals Tom had never even heard of, hoping to stump Tom. He assigned Tom the worst of the grunt work—sorting wet soil with a kitchen sieve—and would complain out loud about the crummy caliber of employees the home office had stuck him with that summer. Tom began having nightmares, calling out in his sleep "I can't see the epidote!" according to his suck-up roommate and co-worker, who would laughingly describe Tom's outbursts to him the next morning. On any given day, Tom didn't know whom he hated more: his boss, for being such an ass, or himself, for failing so miserably at his first job. *Nice, real nice,* Tom found himself thinking. *For this I spent seven years in college.*

Tom had sandwiched a two-week vacation in Seaside between the end of college and the start of the Nevada job. But the atmosphere at home, which had apparently cooled in his absence, had chilled further while he was there. Tom knew Bobbie's relationship with Jim wasn't ideal—she'd admitted as much to Tom in conversations on previous visits. There were money issues, for one thing; Jim wasn't much of a provider, leaving Bobbie to shoulder the bulk of the household bills. Jim often groused about the coastal weather. He had tried to talk Bobbie into moving back to southern California with him, but she wouldn't budge. And there was the matter of Tom's and his siblings' continuing presence in the house. They had all finished high school and left home. So why, Jim complained, were they always hanging around? Showing up, spending the weekend, Tom in particular: fall through spring, Tom would be away at college, but every June he'd show up again, move back into the little cottage on the riverbank. For Tom, it was a simple matter of financial expedience; living rent-free, with three squares a day, he could paint enough houses or catch enough salmon with his brother to cover his expenses for an entire year of college. That, and he liked visiting his mom, and she enjoyed his visits. The two of them just seemed to see the world the same way. Bobbie never gave Tom any indication she wanted him to clear out, to stay somewhere else for the summer. Typically he was working all day anyway. But come suppertime, there they'd be: Tom and occasionally another sibling, his mother, and his stepfather, silent and sullenly waiting out the summer. Jensen himself had left home at the age of fifteen. He just didn't get it, these grown Horning kids always hanging around, crowding the house: pathetic. Some summers Jensen would pick up and head to Eastern Oregon for a summer of rockhounding, returning in September after Tom had returned to Corvallis and the nest was again empty.

The dock had been the last straw—the last of many straws, Tom later realized, but still, he wished it hadn't been his gesture that had tipped the balance. Bobbie had always wanted another dock like the one the tsunami had washed away fifteen years earlier. So Jim had built her one, a floating dock he was pretty proud of, one that rode up and down with the tides. But because it floated, it was never dry; it was always slimy and slick, and no one used it. Tom knew exactly what Bobbie had in mind: a stout pier, rigid, like the dock he'd lain on as a newly arrived five-year-old, staring into the water, watching barnacles wave and shiner perch dart. Bobbie didn't care about taking a boat out; all she wanted was a dry place to sit and watch the bay.

So Tom undertook to start building her such a dock in his short window of time before starting the job in Nevada. He was out in the yard piling up materials for the project—barrels for caissons, old telephone poles—when Jensen approached him one afternoon near the end of his stay. "What are you doing?" his stepfather demanded, uncharacteristically confrontational; usually his criticisms were issued as glancing blows. "What is this?"

"I'm building a new dock for Mom," Tom replied, taken aback. "Like the old one we used to have. For Mom."

Jensen frowned. "You don't know what you're doing," he said, and then he turned and walked off without further comment.

But that wasn't the end of it—Tom could sense it. It was a couple of hours later, the three of them together in the living room watching television, when Jensen turned to Tom and coughed, clearing his throat. "You know, Tom," his stepfather began, tentatively, "you can't—well, at some point, you know, you have to grow up and leave home, become a man. You know what I mean?" Tom just blinked, defensive and dumbfounded. "You need to do like Judy," Jensen continued: "Leave home, make a life for yourself. You can't go on living here with your mother forever."

Forever? I'm home for two weeks, and now it's forever? Tom could sense Bobbie, sitting next to him, tense up, heard her voice, low, quieter than Jensen could hear. "Don't say anything," she warned him under her breath. That in itself was startling, something Bobbie—forthright to a fault—had never done before. That was when Tom knew that this conversation had not come out of the blue but was part of an on-going discussion between the two of them, a conversation about a lot more than Tom and his unsettled living situation, his—as Jensen apparently saw it—extended adolescence. A part of Tom considered walking out at that point, just leaving the room, ignoring Jensen's remark the way he'd ignored previous digs. But he just couldn't do it, not even for Bobbie's pleading. Suddenly the living room seemed charged with electricity and inevitability.

"Oh, really?" Tom replied with studied calm, gathering himself and facing Jensen. "You do know I'm only here for a couple more days, don't you? Going to Nevada? Then to Alaska for the whole summer?

"And come to think of it," he added for good measure, "weren't you living with your mother when you married my mother?"

That scene in the living room—the air thick with tension, his mother's face strained by her effort to keep the peace—came back to Tom months later, as he sat in the back of the helicopter at the end of his last day of work in Alaska. It was late September, and out the chopper's window, Tom could see the blizzards roiling into the Baird Mountains at the western end of the Brooks Range, as if being swept along by the helicopter's own wake: winter already, nipping at his heels. Jensen, meanwhile, was in Needles, California, in the middle of the Mojave Desert—as dry and as different from Seaside, Oregon, as any place in the country. Not an hour after exchanging icy words with Tom in the living room that evening, Jensen had told Bobbie he was leaving Seaside and had asked her to come with him. She had declined. Two days later, Tom left for Nevada, and two days after that, Bobbie had awakened to find Jensen gone. He'd left no note, had just packed some things and driven off before dawn. Almost a week passed before he called to let her know where he'd landed.

Tom's spring and summer of work in Nevada and Alaska had been instructive. He had learned field techniques they don't teach you in school, and he'd learned how to survive a hostile work environment. He wasn't sorry he'd chosen a career in geology, but he was adjusting his sights. Oil exploration—maybe that was the ticket. The price of oil was skyrocketing; it shouldn't be hard to land a job in petroleum exploration. An oil prospector didn't need to be able to see green rocks; he just needed to help his employer see greenbacks. And a master's degree would help broaden his options. A few months in Seaside, long enough to crank out an application for graduate school and to apply for a temporary job in the oil industry, while keeping his mom company and readying the house for winter: that was his plan. From the sound of her letters, Bobbie was still pretty broken up about Jensen's departure: the fact of it, and the way he had left. Tom couldn't help blaming himself, at least in part, for the break-up, but Bobbie insisted it wasn't his fault. Cleary Jensen had been fed up long before Tom had appeared in March. The Horning kids were never going to disappear. Nor was the North Coast weather ever going to improve. Nor was the downward trajectory of Bobbie and Jim's relationship likely to improve, which was—Tom knew—the real crux of the matter.

Tom was a college graduate now, twenty-five years old. Whatever his stepfather thought of him, Tom knew he could take care of himself. He was a geologist, and not a bad one, his color blindness notwithstanding. His summers of painting houses, he hoped, were a thing of the past.

11

The Core Locker

*Use what sense you have, observe well, go ahead and guess your
very best as to what to do. Science is not going to help you much
or prove you wrong until you are very old indeed.*

—Frederick Storrs Baker

MARK DARIENZO WAS LOOKING FOR a quick and easy master's degree: it
was about that simple. All he wanted was to teach geology at a community
college. He had a vision: himself in a tweed jacket with leather elbow
patches, in a cluttered office in an old brick building—or on a modern
college's concrete campus, it really didn't matter—in some out-of-the-way
place. A small department, focused on teaching, no pressure to publish.
He had actually met a guy like that once, back when he was in the Navy
and was taking a field geology course at Bowdoin College in Maine. The
professor was the school's only geologist, a one-man department. He even
wore the tweed jacket. Was that so much to want?

Which is the short version of how Darienzo, at age thirty-five, found
himself squatting among the trampled skunk cabbages and spiky juncus in
the salt marsh at the southern end of Netarts Bay on the northern Oregon
coast, pounding on a meter-long piece of black ABS pipe and plunging it
into the muck. It was a fine day in early spring. The narrow, five-mile-long
sand spit defining the bay's seaward side was densely forested, blocking the
wind and more or less blocking sounds from the state park campground a
quarter mile away. Now and then a car would pass on two-lane Whiskey
Creek Road along the inland edge of the bay. Gulls mewed overhead. A
raven croaked hollowly from its perch in a tall spruce, then relocated in an
ebony flurry to a new perch, and croaked again.

Darienzo had done marsh coring before, back east. Just once, actually—
part of a project for a tidal marsh class at the University of Delaware.
Granted, that was a few years ago. After college, after the Navy and a series

of jobs in New Mexico and Oregon, Darienzo had enrolled the year before in an interdisciplinary master's program at the University of Oregon in Eugene, one that, if he played his cards right, would land him that teaching job in a year or two. His program required him to do some kind of project or thesis, and he'd thought back to that marsh coring in Delaware. What if he did the same thing in Oregon, coring in the salt marsh here? It would tie together the fields he'd chosen for his interdisciplinary degree—geology, physical geography, and oceanography. He would learn field techniques to share with his future students. And it wouldn't cost him much, just a few bucks for some plastic pipe and a roll of Saran Wrap. It would be interesting to compare cores pulled here, on the shore of the Pacific Ocean in Oregon, with what comes up in cores pulled from marshes on the Atlantic coast, though he didn't expect much difference. Continuous peat cores, that's what they were called: tall columns of mucky peat about as deep as you could core. Age-date the layers with radiocarbon dating—old, older, oldest, the deeper you got—and you had the story: a marsh inundated by rising seas, the mean sea level rising slowly and steadily since the end of the last ice age, thousands of years ago. As the sea rose, mud would settle on top of the drowned peat, and new peat would begin growing on top of the old, as vegetation reestablished itself at the edge of the marsh. Continuously. Sand Lake, a shallow bay ringed with marsh on the central coast, seemed like a good spot for such coring. And his advisor had agreed, requesting only that he first do some coring in Netarts Bay, the next bay north. No problem: plenty of marsh at Netarts Bay to core, and only one landowner—the state park—to ask for permission. Simple.

But none of the core samples he had pulled so far from the marsh at the edge of Netarts Bay looked anything like he expected, not one. There were layers of peat in there, definitely, but most of what he was pulling up wasn't plant material at all. What peat he did see was sandwiched among thick layers of mud and even sand, one layer after another: a super-deluxe club sandwich of old marsh vegetation and muck and sand. Not at all what he'd expected.

Darienzo didn't open the cores on site. Instead, he piled the fat, muddy PVC pipes into the back of his green long-bed Toyota pickup and hauled them to Corvallis, to Oregon State University's core lab—a collection of low metal buildings enclosed in cyclone fencing near the edge of campus. There his advisor, OSU marine geologist and oceanographer Curt Peterson, would help him carefully saw each plastic pipe and its contents in half from end to end, opening the filled pipe like a book and revealing its layers in two smooth-faced half-columns. Then they'd wrap each half-core with plastic film and store it among thousands of other half-cores of mud and sand and fossil zooplankton nestled in PVC pipe and stacked on tall shelves in the core locker, a refrigerated warehouse next to the lab where marine geologists and sedimentologists had been archiving core samples pulled from the seafloor for more than forty years.

Darienzo twisted the buried pipe a half-turn and began winching it out of the marsh. The core was stubborn, finally breaking free of the soggy, sucking soil with a reluctant *thwop*.

Wonder what Peterson will have to say about this one, he mused.

PETERSON HAD PLENTY TO SAY about Darienzo's marsh cores—he just hadn't elaborated much yet, not to Darienzo nor anyone else. He knew full well the magnitude of what the buried marsh sequences meant, how they would shake up the geology world's assumptions about the Pacific Northwest coast's past, and its future; he'd been looking for just this kind of evidence for years. This was big—so big that he needed to have his ducks in a row.

"How would you test for coastal records of active plate subduction in Oregon?" was the first question Peterson had been asked back in 1979, in his oral entrance exam for the PhD program at Oregon State. The questioner, oceanography professor Vern Kulm, was a rare bird: one of a handful of scientists who believed that—its recent quiescence notwithstanding—the fault line where the North American and Juan de Fuca plates converged off the Oregon coast was not dead but was continuing to actively subduct. If so, there must have been earthquakes here in the past. And if there were earthquakes, somewhere there was evidence of those quakes.

As early as 1965 Kulm had begun extracting cores from the base of the continental slope off the Oregon coast, painstakingly gouging ten-meter-long samples and winching them onto the deck of the research vessel. He had been surprised to find a remarkable correspondence in the sediment layers found in cores extracted from the ocean floor dozens of miles apart. These layers of olive-green ocean-floor clay interspersed by turbidites—sediments displaced in underwater "landslides" of some kind—seemed to follow the same pattern regardless of where on the seabed they were found, the same thick and thin layers, even similar patterns of burrowing by small sea creatures. Among the materials layered in the cores was Mazama ash from the mountain that now holds Crater Lake; its eruption some 7,700 years ago had blanketed the Columbia River basin with fine ash, much of which was ultimately flushed down streams and rivers to the sea, where it settled on the continental slope and shelf. The ash layer provided Kulm with a crude calendar; anything above the ash layer had settled there less than 7,700 years earlier. But what caused these successive layers, these turbidites, to form across such a broad area about every four to six hundred years? Some kind of very large turbidity currents, clearly. Two likely explanations occurred to Kulm: either very severe storms or periodic earthquakes.

And that was that. The ocean-floor cores went into the core locker at OSU, and the articles Kulm co-authored with his students were shelved. But never quite forgotten.

Kulm himself suspected it was earthquakes. But he had little company, Peterson learned when, within months of starting his PhD program, Peterson volunteered to run the slide projector at a symposium that Kulm organized at Salishan, a resort on the central Oregon coast. Kulm wanted to gauge opinion among his colleagues about what had quietly become one of the burning questions of the day among geologists: Why was the Oregon Subduction Zone—as Kulm called it, though others had begun calling it by the more inclusive name Cascadia Subduction Zone—so quiet, with so few earthquakes? *No* earthquakes, in fact—none of any consequence that anyone knew of. That there was a subduction zone off the Pacific Northwest coast was, by now, accepted; Tanya Atwater and others had established that a decade earlier. But why, when it looked on paper like every other subduction zone around the world, was it aseismic, producing no huge—or even middling, for that matter—earthquakes?

There were, Peterson learned, two primary and opposing viewpoints on that question; he heard both articulated at that conference. One held that the subduction zone had died, that it simply no longer subducted. This view held that Cascadia was once a convergent plate boundary, but it had changed and was now a transform plate boundary like the San Andreas Fault. The Juan de Fuca Plate, it held, now scraped alongside the North American Plate rather than crashing into it and squishing under it. This was the view of nearly every geologist at the meeting.

The alternative view—the so-called "locked zone" theory, argued at the Salishan meeting only by Kulm and Robert Crosson, a geophysicist down from the University of Washington—held that Cascadia was indeed still a convergent boundary, an active subduction zone, just one without a lot of ongoing seismic activity. Maybe the oceanic plate was managing to ease quietly under the continental plate without a lot of fanfare. Or perhaps the plates were so firmly locked, so stuck together, that only very infrequently did the plates unlock in a rare but catastrophic convulsion.

Neither side had convincing evidence—it was still more a matter of opinion than evidence-backed theory. But in the six years since the Salishan meeting—and even earlier, on research cruises with Kulm—Peterson had been doing his best to find that evidence. He had used high-resolution seismic reflection, a technique similar to sonar, to probe the depths of Oregon's Alsea Bay looking for evidence of young—Holocene-era—faults. He had scrutinized sand layers in marsh deposits alongside Alsea Bay, suggesting to his skeptical advisors that the sand layers they contained were deposited by tsunamis. Kulm thought it made sense; the others insisted that the sand was more likely dropped by downriver flooding or incoming storm surges.

Peterson completed his PhD in 1983 but continued the hunt for evidence of old earthquakes along the Cascadia Subduction Zone. Late in 1984 he contacted Tom Ovenshine, who had led the USGS investigations of marsh subsidence in southeastern Alaska resulting from the 1964 earthquake there: the way those salt marshes had dropped in elevation and slowly filled with silt from tides that never quite fully withdrew, the way the tsunami had, in some locations, swept sand into the salt marshes, and the way a new marsh had begun to grow on top of the layers of silt and sand. Oregon's Netarts Bay looked, to Peterson, like the perfect place to look for just such a pattern: unlike every other bay on the Oregon coast, it had no river of any size running into it, so no opportunity for downriver flooding to deposit sediments. And the salt marsh he planned to investigate was fully five miles from the bay mouth; no storm surge was likely to carry ocean sand that far.

Then he got a call from a colleague at the University of Oregon in Eugene: would Peterson mind mentoring a UO grad student with an interest in marsh sedimentology? Sure, he told Darienzo when he called, he'd be happy to help him do some coring at Sand Lake. As long as he'd first pull some cores from the marsh at Netarts Bay.

Now Darienzo was bringing him cores that looked markedly like cores Peterson himself had pulled years earlier from the marshes alongside Alsea Bay, but with even better-defined, clearer sequences of silt and sand and mud and peat. *That's it,* Peterson thought to himself, looking at Darienzo's cores under the fluorescent lights of the core lab at OSU. *We're done! Cascadia mystery solved.* But experience had taught him to take things slowly.

Peterson had witnessed firsthand what can happen when you end up on the wrong side of a paradigm shift in science. As an undergraduate major at San Francisco State University in the early 1970s, Peterson learned that a senior geology faculty member—a micropaleontologist, arguably the most accomplished researcher in the whole department, with fossils named after him, for God's sake—was forced to stop teaching by a committee of students. Why? Because he didn't believe in the new theory of plate tectonics. A student body radicalized and empowered by the anti-war movement was embracing their field's new world view, with no patience for the traditionally plodding pace of change in science. Peterson never forgot that incident, unfolding in the liminal space between paradigms. Science never stands still. But when you're poised at a crossroads, contemplating an abrupt change in direction, he decided, proceed with caution.

PETERSON WAS PRETTY CONFIDENT about what he was seeing in Darienzo's cores, but there was no need to rush; he'd been working toward this moment for six years. First he helped Darienzo buy some additional supplies and sent him back out to Netarts to collect a second round of cores. The second group came back looking just like the first: layers of peat, sand, and silt, a stratigraphic story of abrupt sea level change. Peterson let a week pass before picking up the phone and calling his old boss Ed Clifton at the USGS in Menlo Park. Peterson had worked for Clifton between his undergraduate years at SF State and grad school at Oregon State. It had been Clifton who had urged Peterson to go to OSU: "No one's doing anything up there in your field," Clifton had said. "It's going to be wide open for you."

"I'm really sorry to tell you this," Clifton now ventured after Peterson finished describing what he had seen inside cores pulled from not one but two Oregon salt marshes, at Alsea Bay and Netarts Bay. "There's a fellow up in Washington doing this same kind of research, working out of the Seattle office."

"Oh, really?"

"Yeah. There's a bulletin here with a description of his project. I'll send it to you. You ought to get in touch with him."

As soon as the bulletin arrived at Peterson's office in Corvallis, he sent a copy to Darienzo in Eugene. It sketched out new research projects under way at the USGS. Among them: a project to document what appeared to be—as the report phrased it—coseismic marsh subsidence from an ancient earthquake. The geologist, Brian Atwater (no relation to Tanya) from the agency's Seattle office, had been pulling cores along the southern Washington coast, finding multiple layers of mud and peat that he suggested might be evidence of paleoseismic activity along the Cascadia Subduction Zone. Atwater's preliminary theory: a mega-thrust earthquake on the order of magnitude 8 or 9 had caused a sudden lowering of the land, or subsidence. The lowered land, formerly above the high tide line, became inundated by seawater and subject to tides, which over time settled mud on top of the layers of drowned marsh vegetation—peat—and, in places, sheets of sand deposited by a quake-triggered tsunami sweeping in on the heels of the subsidence.

Darienzo, alone in his apartment in Eugene, had to read the bulletin twice before he could begin to make sense of it. Atwater seemed to be describing the same kinds of things Peterson had talked about when they opened the Netarts Bay cores in the core lab at OSU and peered at their contents. But the bulletin left Darienzo with more questions than answers. His undergraduate degree—a bachelor of arts, not of science—could be described as geology *lite*. He had completed it in 1971, before plate tectonics had yet found a place in college textbooks, though he'd later studied the theory in graduate marine geology courses at the University of Delaware. Earthquakes? His courses hadn't explored them in any detail. Coseismic subsidence—what did that even *mean?* Tsunamis—*here?* And the Cascadia Subduction Zone? Never heard of it.

12

A Prepared Mind

WILLAPA BAY, WASHINGTON, 1986

Come up here and investigate me! Tell me what I am!
—Alfred Wegener, voicing the aurora borealis in a diary entry

BRIAN ATWATER TURNED EAST OFF U.S. HIGHWAY 101 just north of Bay Center, Washington, guiding the '78 Dodge pickup through the sharp twists and dips of South Bend-Palix Road. After a mile or two, the road descended toward what looked like a broad prairie, flat as a runway and nearly treeless, fringed by a dark forest of spruce and hemlock and cedar. As the road leveled, it crossed a narrow river on a wooden bridge surfaced with pocked asphalt. Just beyond the bridge, Atwater veered the truck across the left lane and into a single-car turnout next to a lonely clutch of alders, their spindly trunks a mottled pale gray, their serrated leaves the fresh green of late spring. He and his passenger climbed out and began to loosen the straps that held in place the aluminum canoe jutting fore and aft from atop the truck's cab.

The Niawiakum river wound out of the forest from the east, hugging close to the tree line until it reached the bridge. From there it commenced to meander across the prairie—technically a tidal marsh, Atwater knew, albeit one rarely inundated by summer tides. From his vantage point on the side of the road, Atwater—a tall man with a scruff of beard and a boy's chubby cheeks—could trace the river's westward course toward Willapa Bay for about a quarter-mile, watching it narrow until it disappeared in a bend to the south. Near the bend he spotted what looked like a red-tailed hawk, high in the sky, soaring as it hunted above the even sea of green and yellow grass. Beyond the hawk a half-mile or so were several small, brown lumps: elk, feeding or resting—too far away to tell. The sun rising in the mist to the east turned the tops of the tufted hairgrass to gold and silhouetted a phalanx of a dozen or so Sitka spruce growing in a bunch at the marsh's edge.

At the bottom of the six-foot-deep channel, the narrow river threaded a course across glistening mudflats, running toward the bay in no particular hurry, Atwater noted with a little dismay. In six hours or so, brackish water would be filling the channel to within a foot or two of the bank as the tide surged in. *Should have checked the tide table. Once again.* Now he and Wendy Grant, his research assistant, would have to make a slick slog across the flats and downstream thirty or forty yards to where there was enough water to float a canoe, allowing them to paddle to Atwater's study sites in the marsh. The unhurried water was an opaque brown, with plumes of sediment billowing up through the faster-moving water in the middle of the channel. The top few inches of a spindly, bare tree branch, its other end impaled in the obscured river bottom, nodded in the current, scribing small riffles as the water flowed past it. A single brown spruce needle rode on the surface, near a single struggling bug.

The pair hoisted the canoe and carried it down to the marsh, sliding it easily on the bunchgrass, then began wrestling it down a muddy chute carved in the river bank by previous high tides. The cocoa-colored mud was thick, greasy; walking on it was nearly impossible, like stepping on lard. But Atwater moved through it easily, slowly, with deliberately contained movements. Each of them kept a stabilizing hand on a canoe gunwale.

They and the canoe were thoroughly smeared with mud by the time they reached the water's edge. Grant half-slid, half-walked a few steps back to the bank, retrieving paddles and coring gear that she handed down to Atwater, who loaded them into the canoe. The sun was higher now, the whole marsh alight. They settled into the canoe, Atwater kneeling in the stern, and shoved paddles into the sticky mud to push off toward the middle of the channel. Now their world, right and left and forward, was all brown but for a fringe of dark green above and, overhead, the pale yellow-gray sky. A breeze rattled the alder leaves, but down in the groove of river, the air was still. As she took a paddle stroke, Grant, in the bow, watched as a great blue heron, standing on the bank not far downstream, unfolded her wings in wary anticipation.

ATWATER HAD OWNED THE GRUMMAN CANOE—already banged-up when he bought it at a discount from the dealer in Petaluma, California—since graduate school a decade earlier. Back then he'd never imagined he would someday be paddling it on the Niawiakum River on the Washington coast. As little as a year ago he had not yet set eyes on the Niawiakum. Now he was as familiar with this little river and its surrounding tidelands as he was with any piece of real estate on Earth.

Nine months earlier, in September 1985, Brian Atwater had transferred from the Menlo Park, California, headquarters of the USGS to Seattle, where the University of Washington provided him an office. The move had been motivated by personal reasons. Already parents of a four-year-old daughter, Atwater and his wife had the previous year had a second daughter, born with Down syndrome and requiring sophisticated medical care due to complications she developed later from meningitis. Atwater was then commuting more than an hour each way from their home in San Francisco to work in Menlo Park, and the long commute was straining his family life. In Seattle, he realized, they could afford to buy a house close to the university, where the family would also have access to a specialty children's clinic.

Career-wise, the move's timing turned out to be fortuitous. In the 1970s the USGS had embarked on an effort to thoroughly study earthquake hazards in a series of metropolitan areas beginning with San Francisco and followed by Los Angeles and Salt Lake City. Now it was Seattle's turn. In October, less than a month after Atwater had moved north, the agency held a big kick-off celebration in the ballroom of a downtown Seattle hotel. Among the speakers was Tom Heaton, a USGS seismologist from Pasadena, California.

Atwater had heard Heaton speak once before, back in Menlo Park. Heaton was a big name at the USGS, a guy with big ideas, a big head of curly blond hair, the gravelly baritone of a country western singer, and an infectious enthusiasm for his subject. That night in Seattle he spoke about, among other things, the magnitude 8 earthquake that had struck Mexico City just weeks earlier, on September 19, 1985, toppling ancient ruins and modern office towers and hospitals and houses and bringing the city and its inhabitants to their knees. Like Seattle, he noted, Mexico City sat atop a subduction zone, an area where two tectonic plates converge. The trench where the fault had ruptured was more than two hundred miles away from the city—about the same distance from Seattle to the leading edge of the Cascadia Subduction Zone. Mexico City, like many cities, was built in a sedimentary basin. Scientists suspected that the extensive damage suffered in Mexico City might have been a function of long-period earthquake waves moving easily through, and getting amplified within, that basin.

Atwater had a background in Quaternary geology—*young* geology—and was vaguely familiar with the phenomenon of liquefaction: the way loosely consolidated sediments like sand and dirt can turn to soft, wet muck during an earthquake, causing buildings and roads built upon it to collapse. He had already decided his contribution to the USGS quake hazard effort in Seattle would be to make liquefaction maps of the Seattle area, identifying the places where the ground was most likely to lose strength in the event of a big quake.

But Heaton had other interests that he addressed at the meeting. In the early 1970s, the Washington Public Power Supply System had proposed construction of three nuclear reactors at Hanford in Eastern Washington and two at Satsop, twenty miles east of Grays Harbor on the southern Washington coast. By the early 1980s, the Satsop project had become a financial and political debacle, and construction of one Satsop plant had been terminated. In 1983 WPPSS defaulted on $2.25 billion in bonds. Bondholders sued. But WPPSS officials still held out hope that construction on the other plant at Satsop, then three-quarters finished, might resume.

Meanwhile, the Nuclear Regulatory Commission had begun pressing seismologists about the earthquake potential of this Cascadia Subduction Zone they'd been hearing about. Was the southern Washington coast in fact a dangerous place to build such a plant? Oregon's sole nuclear power plant, Trojan, stood about fifty miles upstream from the mouth of the Columbia River and was in full operation.* There were no nuclear power plants in British Columbia; to the south, the closest one was near Eureka, California, but the Humboldt Bay Power Plant had been shut down in 1976 and was on its way to being decommissioned. Heaton was among those scientists who had been working to assess Cascadia's potential for quakes, mainly by comparing it to subduction zones where great earthquakes had occurred in the twentieth century. Heaton's position, at that point, could best be expressed in the negative: there was nothing he had found that clearly distinguished Cascadia from other subduction zones. Lacking evidence of large earthquakes along the Pacific Northwest coast in the past, he said, one couldn't conclude that Cascadia would produce a large earthquake in the future. But one should *not* conclude that it was *incapable* of producing one.

---

*Trojan Nuclear Power plant was ultimately shut down in 1992 and its cooling tower demolished in 2006.

ATWATER DIDN'T KNOW MUCH ABOUT the Cascadia Subduction Zone. It hadn't been discussed at Stanford, where he had studied geology as an undergraduate, nor at the University of Delaware, which awarded him his PhD. It wasn't a huge focus at the USGS; of greater interest quake-wise were the many known and active fault zones around the country. He considered himself an agnostic on the question of Cascadia's seismic potential. He'd read the work of George Plafker, Tom Ovenshine, and others describing the effects of the 1964 Alaska quake on the landscape up there. What if, in addition to the liquefaction mapping in Seattle, he did a little poking around on the Washington coast, looking for similar clues in the landscape there?

Then, just four months after the move to Seattle, Atwater's twenty-month-old daughter, Sarah, died. Her needs had kept him close to home, ready to help with her care and respond to emergencies. It would now be possible to consider making those trips to the outer coast, a two or three hour's drive away.

Atwater was familiar with a seminal 1979 paper by Masataka Ando and Emery Balazs, two scientists seeking evidence that the Juan de Fuca Plate was indeed subducting under the North American Plate. In Japan, geodetic measurements indicated that the coastline adjacent to the Japan Trench—where the Pacific Plate was apparently subducting under an arm of the North American Plate—was being dragged down, its elevation lowering, between large earthquakes. If Cascadia were an active subduction zone, also poised between earthquakes, they figured, they should find evidence that the Pacific Northwest coast was being pulled down in a similar manner. So they did what scientists did before the days of the Global Positioning System: they examined historical trends at tide gauges wherever they could find them on the Northwest coast, to determine whether the sea level was rising in relation to the land. And they scrutinized the results of the National Geodetic Survey's "re-leveling" of geodetic benchmarks—bronze medallions placed on bedrock—in western Washington, precisely measuring the benchmarks' elevation repeatedly over decades. To their surprise, they found that Washington state's coastline was *not* being dragged down, but neither was it staying level; apparently it was *rising* in relation to the interior. Meanwhile, the elevation of Puget Sound had apparently not changed at all. Their conclusion: since the leading edge of the continent was not getting dragged down by the convergence of two plates, then the plates must not be locked, must not be building up strain that would be released in an earthquake. Rather, the land-level increase they observed must be part of a continuous, permanent uplifting of the land, not a temporary, elastic deformation that would spring back after an earthquake.

That study would be his starting point, Atwater decided. He would look for geological evidence—clues in the mud—indicating that the Washington shoreline was going up or down, to compare with Ando and Balazs's geodetic evidence. But the trend line the two scientists had created spanned only seventy years. Atwater, taking marsh core samples, could time-travel much further back. He should be able to compare the age of marshes on Puget Sound with the age of marshes on the outer coast over thousands of years, as the seas rose and the marshes attempted to keep their heads above water. If Ando and Balazs were right, the three-thousand-year-old shoreline should be thirty feet higher at, say, Neah Bay, on the outer coast, than at Whidbey Island in Puget Sound, Atwater calculated. He figured he shouldn't have much trouble detecting that magnitude of difference. Neah Bay was an obvious place to sample for the outer coast, due to the presence of a tide gauge that had continuously measured sea level relative to the land since the mid-1930s.

The town of Neah Bay is tucked in a notch of shoreline on the Strait of Juan de Fuca, the wide sleeve of saltwater separating Washington State from Canada's Vancouver Island. It sits just east of Cape Flattery, the knob of rock jutting into the Pacific Ocean at the northwesternmost corner of Washington. The cape itself is a rocky, forested upland; immediately to the southeast the land rises toward the Olympic Mountains. Between the Olympics and the cape lies a broad, flat lowland known as Waatch Prairie, stretching three miles from Neah Bay to Hobuck Beach, on the Pacific Ocean side of the peninsula. With the permission of the Makah Tribal Council—representatives of the Indians whose ancestors had inhabited this area for millennia—Atwater wandered out onto the prairie on a rainy morning in March 1986 to have a look around.

The prairie itself wasn't low enough to be inundated by high tide, but it would yield soil samples that could be age-dated and compared with the Whidbey Island samples. He spent most of the day pushing a gouge corer into soil here and there. Then in late afternoon, an hour or two before dark, Atwater got back in his pickup truck and followed the road from Neah Bay out to Hobuck Beach, looked around a bit, then backtracked nearly to Neah Bay and parked along the road.

He got out of the truck and, pushing his way through wild roses and bare-leafed willows, made his way down to the Waatch River, a shallow stream that drains the prairie, and followed it upstream a few hundred yards. Here the river makes a series of tight meanders, as if uncertain what direction the ocean is. On a hunch, he dropped over the edge and squatted next to the water, peering at the cut bank.

Below the mat of typical salt marsh plants at the top of the bank was a layer of taupe-colored mud and sand and, below that, a darker layer of what appeared to be old marsh. Animals had burrowed tunnels through all the layers, but the buried soil, below the high-tide line, was clearly too deep to support plant life. Atwater stepped back, considering. He was familiar with continuous peat cores—tangible evidence of a salt marsh that had kept up with rising sea levels over thousands of years. He'd seen them—had pulled such cores himself—in coastal marshes in the mid-Atlantic. He'd seen many such cores firsthand when he'd paddled his Grumman canoe into the tule marshes in the upper reaches of San Francisco Bay, where the bay stretches narrow fingers deep into California's Central Valley, while conducting research for his PhD dissertation. Which made the layers in this creek bank on the northern Washington coast stick out like a sore thumb. Atwater knew what gradual sea level rise looked like, and this wasn't it. For some reason, at some point in the past, this marsh must have suddenly subsided and been covered with saltwater flooding in from the ocean. So Ando and Balazs were wrong, he realized: the Pacific coast had *not* been rising continuously, relentlessly, permanently. Standing in the cold drizzle of that late March afternoon, Atwater couldn't help but smile.

Not for a day or two did Atwater think to send the postcard. He was still in Neah Bay, and it was still raining. After a visit with tribal authorities at their office in town, Atwater had popped into Washburn's Grocery, choosing a picture postcard from the selection in the rack. Then he went outside and around the corner, into the post office attached to one side of the store. There he bought a stamp and inquired about the zip code for Pasadena, California. Atwater had never met Tom Heaton face to face, but he was pretty sure Heaton would be as interested in the evidence of land-level change he'd stumbled across as he was himself.

Was geology about to provide the smoking gun, resolving the Cascadia question? It was an exciting prospect. But the more relevant question at the moment, Atwater knew, was a more pragmatic one: What would be the best place to go next to look for evidence of earthquake subsidence, and how soon could he get there?

WHERE HE DECIDED TO GO WAS TO the big tidal marshes on Washington's southwest coast, where he planned one-day reconnaissance visits to Willapa Bay, Grays Harbor, the lower Columbia River, and the lower Copalis River. At each place, invariably he found buried marsh soils much like those at Neah Bay, repeated layers: evidence of what he began to think of as "jerky submergence" events. And at each site, it seemed, he found some new wrinkle. Sand layers atop buried marsh soils, for instance: evidence of tsunami surges rolling in and dumping sand on top of the suddenly sunken coastline. Root wads from buried stumps preserved in the cold mud, presumably killed when the trees' roots had been suddenly plunged into saltwater by a coastline jerkily submerging. Souvenirs of an event Atwater began to suspect may have happened not all that long ago. *Ankarty*, in the Makah language, for by this time he had heard from someone—maybe it was Tom Heaton—a Makah Indian tale that white schoolteacher James Swan had heard in Neah Bay in 1864 and recorded in his diary. Some time in the not so distant past, a Makah elder had reported, the sea had flowed in and flooded this very lowland, Waatch Prairie, cutting off the cape from the mainland, killing people and wrecking canoes all along the coast. As to when this had occurred, the elder said *ankarty*: "not a very remote period."

By late spring, Atwater decided to focus his efforts for the 1986 field season on Willapa Bay and the little tidal river that entered the bay almost directly east of the bay mouth at the northern tip of the Long Beach Peninsula. The marshes in the Niawiakum River valley had not been diked by farmers; they had hardly been disturbed by anyone. U.S. Highway 101 crossed the river near its confluence with the bay, and the old highway—now South Bend-Palix Road—crossed it a few miles upstream, giving Atwater two convenient access points. At first he visited alone, but as the fieldwork geared up, he recruited Wendy Grant, a seismologist also with the USGS in Seattle, to help. Back in the 1970s, Ed Clifton and his group from the USGS in Menlo Park had stayed in the KOA campground in nearby Bay Center while they investigated the uplifted marine terraces along the Willapa Bay shoreline. Now Atwater and Grant made it their headquarters as well, sleeping in tents. A little unpainted aluminum trailer—a '53 Westwood—that Atwater had previously used for fieldwork in eastern Washington functioned as their kitchen and office.

The task Atwater set for himself was to pull core samples at sites all along the lower Niawiakum, looking for the kind of stratigraphic motif he'd seen in Neah Bay and elsewhere along the outer coast. To look for it, and to *not* look for it: in science, Atwater had found, once you see a thing, or once you know what to look for, you're likely to find it. It's not quite the same thing as bias: *a prepared mind for the chance* was how he thought of it. Now that he knew what he was looking for, he started looking for the *other*: continuous peat cores, the kind of thing he had seen along the Atlantic coast and the coast of central and southern California—coasts without a subduction zone offshore. Such a find would pose an argument *against* suddenness, against jerky submergence, and that would be a valuable find—the alternative view. If he did find continuous peat cores, he'd have to reconsider the scenario he'd been forming in his mind: a very episodic but very violent history of earthquakes along the Pacific Northwest coast.

He never found any continuous peat cores.

But he did find a better way to look at stratigraphy, one that quickly became his default technique. It was on a show-and-tell trip down the Niawiakum with a group of visiting scientists when, once again, he cursed himself for not paying closer attention to the tides. There he was, putting in at low tide, meaning a long march across slick mudflats to get to his study sites. The flotilla started drifting downstream, down the narrow, mud-walled canyon, with nothing to see but mud. And quite suddenly, Atwater found himself looking at that muddy bank with fresh eyes. There it was, his sample: thousands of years of stratigraphy in the riverbank itself, requiring nothing more than a little digging to clear away the outer coating of tidal mud. Tide was time—the lower the tide, the more riverbank he could see, and the further back in time he could look. The exposed riverbank provided a window on the past, one not merely an inch wide, such as his core barrel offered, but as wide as the river was long.

13

A Fresh Sandbox

Humboldt County, California, 1986

A scientist must also be absolutely like a child. If he sees a thing, he must say that he sees it, whether it was what he thought he was going to see or not. See first, think later, then test. But always see first. Otherwise you will only see what you were expecting.
—Douglas Adams, *So Long, and Thanks for All the Fish*

Pushing aside the brambles at the edge of Canyon Creek, Gary Carver and Tom Stephens scrambled up the muddy incline to the edge of the gravel road, pausing at the top to catch their breath and reorient. It had been a couple of hours since Tom had parked the blue jeep some miles back at the gate and they had taken off on foot, following logging roads through private timberlands above Korbel, northeast of Eureka, California. Here the forest was second growth; most of the really big trees, the Douglas firs and redwoods, had long since been cut. But trees were not what interested Carver and Stephens. What they were seeking were exposures.

If you're a geologist looking for rocks in the backwoods of Humboldt County, California, where the rain falls steadily six months of the year and even the summers tend to be cool and foggy, and the soil is black and rich and overlaid with spongy humus, and the rhododendrons and huckleberry bushes form a dense understory taller than your head, an understory dwarfed by trees that, full grown, are among the tallest in the world, you either need some heavy equipment to trench down under the dirt, or you need to find spots where someone else has already done the work for you. Road cuts, for instance: places where a road grader's blade has gouged out a hillside to build a road for log trucks to travel and, in that process, has exposed a neat slice of the Earth's history. Gary and Tom had dug plenty of trenches in the pastures and woods above Humboldt Bay in the past decade. But you can't dig trenches everywhere. Sometimes you can make more progress by simply getting out and looking for exposures.

Creek beds were another source of exposures. Over time, water sluicing down ravines carves away dirt and sand, exposing the harder rock beneath. From out of his daypack, Gary—long-legged and long-armed, with long hair and a short, scraggly beard and a red kerchief around his neck—now pulled out a well-worn Simpson Timber Company map and partially unfolded it, revealing a vast lacework of logging roads and natural features. Canyon Creek, for one, which he now traced with a finger, following the pair's presumptive route from one road to the next, then along the creek for some distance and back up to the road, to the point where they now found themselves. Gary refolded the map and stuffed it back inside his pack, and he and Tom took off down the road, Gary's neck craning forward as if on the scent, Tom's compact athleticism and his hickory walking stick keeping pace with his companion's long strides. It wasn't really a walking stick per se: more like a long-handled geology pick, with a World War II-vintage trenching tool set atop a long wooden handle that Tom himself had carved by hand. Nice for catching one's balance on scree slopes; ideal for gouging and scraping the soft, young rocks Tom habitually found himself scrutinizing.

The two had been friends and geological co-conspirators since 1973, when Gary first arrived in Arcata to join the geology faculty at Humboldt State University. Tom was a graduating senior in the HSU geology department and was already working for the USGS, mapping creek basins in the surrounding redwood forests. Two hippie geologists with a mutual scorn for convention and a shared passion for steelhead fishing and getting lost in the woods in the name of science: theirs was a natural partnership.

What they were seeking in those exposures was evidence of faults: displacement of land caused by the movement of tectonic plates. Young, active faults in particular. The search had begun serendipitously back in 1976, when Tom was mapping in the hills north of Trinidad Head just east of Big Lagoon, a two-mile-long teardrop of brackish water separated from the ocean by a slender sandbar. There in the brushy foothills he stumbled across an outcropping with an unusual fracture pattern that neither he nor Gary—whom he immediately brought to have a look—had ever seen before: microfractures in a distinctive rhombohedral pattern, like hundreds of little diamonds, inches apart. And it was in unconsolidated rock: marine sediment so young, it had been uplifted from the seafloor before it had even had time to compress into standstone by the weight of the ocean. Not only that, but something—movement, a fault of some kind—had later fractured that not-yet-rock. The outcrop was telling a story, but one that seemed—to Tom and Gary—almost more like a tall tale.

Hardly any geologic mapping had ever been done in the Humboldt Bay region—nothing at all since the 1940s and '50s. It was, Tom liked to say, a fresh sandbox. Earlier geologists had found evidence of faults around Humboldt Bay, but they believed them to be mostly small, and old—inactive. Nothing had changed since the advent of plate tectonics; the faults in Humboldt were assumed to be of the San Andreas variety: strike-slip, from two plates sliding alongside one another. But the curiously fractured sediments Tom and Gary began finding in dozens of sites suggested a very different story: a convergence of tectonic plates, and not very long ago.

The hunt was on. Gary talked Simpson Timber Company into loaning him keys to the gates that kept the public out of the company's vast holdings of forestlands east of U.S. Highway 101. In the late 1970s Gary and Tom secured an Earthquake Hazard Reduction Grant from the California Division of Mines and Geology and used it to locate and map a half-dozen faults, a couple of them very large, between Trinidad Head and Cape Mendocino—the Mad River Fault Zone, they came to call it. Gary was convinced that these were young faults. But how young? In some sediments, he and Tom found fossil seashells of a species that had been around for three million years. In less than three million years, then, these sediments had been uplifted, twisted into big folds, and cut by faults, big faults. But three million years was still, even to a geologist, a pretty big window.

Up ahead the road made a broad curve, and as they rounded the bend, Gary and Tom were pleased to see a road cut ahead. And then both men stopped in their tracks, and stared.

It was more than a road cut: it was the back wall of a quarry, where fill material had apparently been mined for road construction. Even from fifty yards, they could see that this cut had something they had never seen before in all of Humboldt County, blazing across the quarry's vertical wall: a layer of bright white rock nearly four feet thick.

Gary and Tom looked wordlessly at one another, disbelieving. Then, with a kind of rapturous restraint, they walked to the wall and slid their fingers across it, feeling the grit. Even before touching the white rock, they knew what it was.

Gary *whooped* and swung around to high-five Tom, who was alternately shaking his head *no* in disbelief and nodding *yes* and grinning like a madman. Then Tom swung at the quarry wall with the business end of his long-handled pick, dislodging a huge chunk that tumbled toward him. He reflexively raised the stick to deflect the blow, and—*crack!*—the handle of his treasured handmade pick snapped in two.

Tom didn't even care. What he and Gary had found was, for their purposes, nothing less than the Holy Grail.

It was volcanic ash, clearly. And to a geologist, ash equals time. Each prehistoric ashfall has a unique geochemical fingerprint linking it to a particular volcanic eruption. They would need to send a sample to the USGS for testing, but they already had a hunch that it was Huckleberry Ridge tuff, a product of one of the largest of three great eruptions of the Yellowstone hotspot some 2.1 million years ago. The ash that composed this tuff had shot out of a volcanic vent a thousand miles away, had drifted on the wind, and fallen on the near-shore ocean, filtering down through the water column and settling thick and spongy on the seafloor. Over time it had been covered with sand and other sediments. The ash, semi-consolidated along with the marine sediments above and below it, had been uplifted and folded and, eventually, broken during earthquakes in long a series of tectonic events that Tom and Gary could now see spinning out practically in front of their eyes.

Adding to what they knew about the origin of marine terraces and the chronology of sea level change, it was now clear that some of the faults they'd identified had cut across sediments laid down no more than ten thousand years ago—the Holocene epoch, *now* in geologic terms. That made the faults they had found, by definition, active faults.

They stuffed several chunks of ash into their daypacks. Tom gathered up the broken pieces of pick handle. Then they turned around and began the walk back to the jeep, where Tom had tucked their stash inside the first aid kit in the glove box.

It was time to celebrate.

CHAIRLIFTS WERE STILL A NOVELTY in the 1940s and '50s, when Gary Carver was growing up in the sleepy little mountain communities of Reno and South Lake Tahoe. As a kid he flyfished and ran a trapline, but by high school, nothing eclipsed his passion for skiing. He stumbled through his first year and a half at the University of Nevada Reno, where he was a civil engineering major. The day of the first major snowfall, he took a day off to go skiing and never went back, not even to withdraw. He stayed on the mountain, working and skiing, the rest of that year and the two after that. Then his life as a ski bum came to a crashing halt in spring 1964. On April 15, at the end of his 123rd consecutive day of skiing, Carver climbed on a bus headed down the mountain toward Fort Ord and, presumably, Vietnam.

From Fort Ord, he was transferred to Fort Lewis. The Army was not a good fit for Carver; his strategy was to screw up enough in basic training that he would eventually be kicked out. Eventually it worked; a sympathetic sergeant (who probably didn't want to go into combat with the likes of Carver) helped him transfer into the National Guard. The University of Nevada decided to give him another chance. By then Carver knew that engineering wasn't his thing. On a whim, he signed up for a geology class.

It was as if someone had thrown a switch. He was back in the mountains, among the rocks and glaciers he so knew and loved, learning the names of the minerals and processes that composed those rocks, those rivers of ice. He promptly quit skiing and finished the bachelor's degree he had begun eight years earlier. One year later he had his master's degree. Two and a half years after that he completed his PhD.

Carver's episodic higher education years—1960 to 1972—roughly spanned the most turbulent period in the history of the earth sciences. It was while he was pursuing his master's degree at the UNR's Mackay School of Mines in spring 1968 that Carver first read a comprehensive rendering of the theory of plate tectonics, though it wasn't yet called that. Late one day, heading home, he ran into his supervising professor, Bert Slemmons, scurrying down a hallway and clutching an armload of papers, exclaiming, "Read this! You've got to read this, it's revolutionary!" and handing a copy to Carver and every grad student and faculty member he encountered. It was an article slated for publication in the *Journal of Geophysical Research* and written by colleagues of Slemmons, seismologists at Columbia University. As he handed out photocopies of "Seismology and the new global tectonics," Slemmons summoned Carver and the others outside to read it together. There in the late afternoon Nevada sun, at the feet of a bronze likeness of Comstock Lode "Bonanza King" John Mackay himself, Carver read while Slemmons annotated the piece out loud, spelling out its significance, paragraph by paragraph. It was, for Carver and probably everyone else sprawled on the steps at the entrance to the School of Mines that afternoon, a turning point. All that was left, it seemed, was for Carver and his generation of young geologists to figure out the details.

But plate tectonics was not the only revolution under way. One year later, in 1970, Carver was at the University of Washington, starting work on his PhD and helping lead the students' Strike Coalition Steering Committee. News of the killing of four unarmed students by members of the Ohio National Guard at Kent State University on May 4 had fueled the already white-hot anti-war movement nationwide. In Seattle, Carver was among those who, in protest to the war, managed to effectively shut down the UW campus for a week. In a month of explosive protests, on campus and off, he developed the habit of wearing a bandana around his neck. It could provide protection when the police threw tear gas at demonstrators. And it could be employed as a sling to lob hot tear gas canisters back at police.

The bandana became his signature accessory. He was wearing it the day in 1971 when, fulfilling his duties as president of the university's Geology Club, he drove to Sea-Tac airport to pick up Tanya Atwater, then a post-doc at Scripps Institute of Oceanography and a rising star in the earth sciences revolution. Her talk at the Asilomar conference in December 1969 and the paper that followed, "Implications of plate tectonics for the Cenozoic tectonic evolution of western North America," had taken earth sciences by storm, and invitations to speak had poured in from all over the West. That evening in the auditorium at the University of Washington, Atwater, her mop of curly hair untethered, proceeded to explain her version of how the world as we know it, the oceans and continents, had come into being and where they were headed. And when a gray-bearded faculty member not yet on board with this new-fangled plate tectonics business raised a hand to challenge her conclusions, Atwater politely but firmly pointed out the flaws in his reasoning and the gaps in his knowledge of current research. That professor did not attend the faculty-graduate student reception after her talk that evening. Much less the after-party Atwater and Carver and his friends made their way to later that night. Off-campus.

It had been Tanya Atwater's elegant thesis that had put together the whole tectonic history of western North America for Gary Carver back in 1970—how the San Andreas Fault and the Mendocino Fault and the Cascadia Subduction Zone converged at the Mendocino Triple Junction off northern California's Cape Mendocino. Now, a decade and a half later, it was to that history that Professor Carver's thoughts strayed. The San Andreas Fault, running down the middle of California, had been thoroughly studied. But the Cascadia Subduction Zone, out on the continental slope, had remained inscrutable. It was especially hard to get a handle on it down here at the southern end, where the Gorda Plate, a southern fragment of the Juan de Fuca Plate, seemed to be pushing under the North American Plate right next to where the North American Plate was sliding alongside the Pacific Plate. The continental faults Gary and Tom had found and named gave every indication of being young, active faults. So why were they so quiet? And the Cascadia Subduction Zone itself—could it really be aseismic, as most of the scientists with the USGS seemed to think?

Then Carver met Sam Clarke.

It was spring 1986, at a regional meeting of the Geological Society of America in Los Angeles. Carver's presentation on the onshore thrust faults around Humboldt Bay went fairly well, he thought, given that he himself was still uncertain about what it all meant. The following day, on a whim, he decided to listen in on a promising-sounding presentation by a USGS marine geologist.

Sam Clarke, Carver learned, had been working from a boat above the continental shelf north of Cape Mendocino, using acoustic reflection techniques to map the seafloor. After explaining his research methods, Clarke clicked to a slide with a map of his findings.

Carver was flabbergasted. Apparently he and Clarke had been simultaneously working different ends of the same piece of real estate, Carver on the continent and Clarke on the continental shelf, and had come up with essentially the same story. Clarke's seafloor deformations appeared to be direct extensions of the fault lines Carver had been tracking on land, point for point.

Clarke left the conference center before Carver had a chance to corner him, but it wasn't hard to track him to the hotel where he was staying. Carver returned to his own room to fetch his map.

"Hi, I'm Gary Carver," Carver began when Clarke opened his hotel room door. "I teach geology at Humboldt State, and I want to show you my map." Clarke had missed Carver's presentation; he didn't know what this tall, wild-eyed, long-haired geologist with the red bandana was talking about. But Clarke dutifully unrolled his map on the hotel bed while Carver unfurled his, both of them hand-drawn in the style of the day: Rapidograph pen and press-on lettering on mylar film, barbed lines indicating the location of fault lines. They butted the two maps together.

By coincidence, both geologists had mapped using the same scale, 1:250,000. But it wouldn't have mattered. Even at a glance, the correspondence was remarkable. The little black fault lines Carver had drawn to their terminus at the shoreline were picked up and extended on Clarke's map of undersea deformations. Those deformations started at the shoreline and trended northwest across the reach of the continental shelf. Clarke, working from a boat, didn't and couldn't do the kind of looking and touching and age-dating of the undersea faults that Carver and Stephens had done with the faults they found on land. But extrapolating from Carver's data, it wasn't hard for Clarke to now guess the genesis of the ridges and troughs he'd found there. It was as if each geologist had been reading different halves of the same novel: curious in isolation, and not revelatory until the two halves were joined.

This chance meeting of the maps was all the evidence Carver needed. He could see now that these faults were expressions of a subduction zone that had indeed made huge, sudden adjustments in the geologically recent past, adjustments of the kind commonly known as earthquakes: big ones. There was nothing dead about the Cascadia Subduction Zone, he was now convinced.

This, Carver mused, *is going to blow some minds.*

14

Last Glance

TOM, 1987

TOM WAS ALONE IN THE DUPLEX a couple of miles north of the Oregon State University campus that he shared with a fellow geology grad student, wrestling with a paragraph in his thesis that didn't seem to want to come out right, when he got the call Saturday morning, January 17. It was his sister Judy in Portland.

Bobbie was dead, Judy told Tom. Jan, David's wife, had found her in her bed that morning, gone.

It was the call Tom had been half-expecting every day, the call a part of him didn't really believe would ever come. Tom had talked with her just a few days earlier—*NOVA*, Tom's favorite TV show, had been airing a program about the mysteries of embryology, and before he'd even had a chance to pick up the phone to call Bobbie, she had called him.

"Why don't you come down for a visit," she had said before hanging up. "You ought to come down." But he had been busy. The pressure was on at OSU: he'd been in and out of grad school for nearly six years, and there was a seven-year limit on completing a master's thesis. He had labs to run, and every time he finished another section of the writing, it seemed like there was even more, not less, yet to do. Besides, he had been up to see her just three weeks earlier, over winter break. "I'll try to get down in a couple of weeks," he'd told her. Now he was kicking himself. If only he'd listened.

It had been fall 1982, and Tom was two years into graduate school at Oregon State, when the first red flag went up. Tom had spent the summer working for mining conglomerate ASARCO, prospecting for gold in the Rocky Mountains. The eight months he'd spent with Unocal in Anchorage the summer before starting grad school—sitting in an office tower, scrutinizing petroleum drill records—had left him a little flat, and he had wanted to take another stab at field geology. He had spent the next summer in Peru, collecting data for his master's thesis, and the experience had boosted his confidence in the field. He'd gotten better at using clues other than color to identify rocks (and, just to be sure, doing plenty of geochemical sampling).

But after two years of grad school, he was considering taking a break and going back to work in the oil industry for a while. The price of oil was at a historical high. Three years earlier it had spiked to $38 a barrel, from barely $3 a barrel a decade earlier; it had slid some since then, but not by much. Oil companies were scooping up geologists by the handfuls, wining and dining them—actually, it was usually beer and pizza feeds—to entice them into jobs. Tom had received eight job offers that fall, had accepted interviewing trips to San Francisco and Denver at the companies' expense. It was more than a little heady, and Tom had half a mind to put his master's degree on hold for a while and take a job with Chevron in San Francisco. Then in early December, less than a week after he'd returned to Corvallis from his interview with Chevron, he got a phone call from Judy. Bobbie was in an ambulance, she said, on her way to Portland for surgery—if she survived the trip.

A career nurse who by then had her own home health business, Bobbie had known that the tenderness she'd been feeling in her chest was not something to ignore, as she later told Tom. She'd driven herself to Seaside Hospital, where imaging tests showed a bulge in her aortic arch. In Portland, surgeons had opened her chest and replaced the weakened section of aorta with a synthetic graft. Tom called the Chevron recruiter with his regrets; it didn't seem like the right time to move to San Francisco. That Christmas, for the first time in twenty-three years, the Horning kids didn't celebrate the holiday at the family house in Seaside but gathered at Judy's house in Portland, where Bobbie was recuperating.

At the end of spring term, Tom headed to Seaside for the summer. Bobbie was back on her feet by then but skinnier, less energetic. She'd developed, for the first time in her life, a fondness for television. When he wasn't busy painting houses, Tom pitched in to help his siblings and their spouses dismantle Jim Jensen's little house and carve out, at Bobbie's request, a small garden plot, a place for her to putter just steps from the bay.

By the following spring, Bobbie seemed fully recovered and settled into a quiet retirement. With copper and gold prices down and oil company jobs harder to come by, Tom took a new tack for his summer employment. He hired on as a structural geologist with the geothermal branch of Unocal, exploring the geothermal fields of northern California and helping the company assess the potential for making money out of the heat of the Earth. He returned to northern California the following summer and was there, seeking steam and dodging pot farmers and poison oak, when he got word that Bobbie was going in for a second surgery. This time the aneurysm was in her abdomen.

Technically the surgery was successful. But her aorta was continuing to wear out, to "delaminate," as the doctors described it, and there wasn't anything they could do to stop that process. When, less than a year later, she was diagnosed with yet a third aortic aneurysm, Bobbie opted against more surgery. She knew that her odds of surviving another operation weren't good, that even if she did survive, her quality of life would probably be so poor that she would wish she hadn't. Instead, she settled in at the house she loved, the house at the bay mouth where she had lived and raised her kids, where for a quarter-century she had watched the tides ebb and flow, the seasons change. And she began, with characteristic stoicism and a kind of grudging equanimity, to accept the fact that she was dying. Tom had spent that next summer close to home, bouncing between Corvallis and Seaside, taking odd jobs to cover his rent and lab expenses, working on his thesis, checking in on Bobbie.

Christmas 1986 had been rough. All the Horning kids knew Bobbie didn't have long to live, but Tom was the only one who had come home for Christmas. Tom tried not to be bitter, or self-righteous, or at least not to show it. All four of his siblings were by now married, with spouses and children and in-laws and other holiday obligations beyond Bobbie Horning; he understood that. Still, it annoyed Tom; surely they all knew it might be their mother's last Christmas. It annoyed him that no one else had managed to visit her over the holiday. It annoyed him that his brother had bought Bobbie such a pathetic little Christmas tree—even though she'd asked for a small tree. It annoyed him that Ronald Reagan had allowed his people to screw things up so badly. Bobbie had taken to leaving the TV on, and the non-stop news about the unfolding Iran-Contra scandal had become a kind of irritating Greek chorus. "Mistakes were made," President Reagan had said. *That's for sure,* Tom groused. Tom was cranky all Christmas, carping about politics, about the state of the world, and peeved at himself for being so negative at such a time. In the back of his mind, he knew that President Reagan wasn't his biggest problem. It was something else, something bigger than world politics, something he couldn't fix, so close to him that he didn't even have words to talk about it.

And then, just three weeks later, the call from Judy. That Bobbie had been alone: that was the worst part. Not the fact of her death—that was inevitable—but that she'd died alone in her bed, eyes wide open, according to Jan. Probably frightened, maybe in pain, unable to get out of bed, to call anyone, for who knew how long. That Tom had not been there, had not picked up on her cues, hadn't shown up nor called in the family to comfort her in her final hours as she had so often comforted them as kids, as young adults, and comforted myriad others in the course of her nursing career.

She had been widowed young, had rallied and moved on. She'd worked hard all her life, supporting herself and her five kids and helping to carry two husbands. She had taken care of nearly everyone in town, at one time or another: as the RN in a general practitioner's office, as head of nursing at the hospital, and ultimately as a hospice nurse, caring for dying neighbors. She'd survived breast cancer and twenty-seven years of winter storms and even a tsunami from Alaska in a house perched at the edge of the sea. She was so young, relatively speaking; she'd barely had time to enjoy her retirement. She didn't get to collect even one Social Security check, dying a month shy of her sixty-fifth birthday.

In her will, Bobbie had left the house to all five of her children. David and Jan, then living in a rented house in Seaside, agreed to move in to keep the place occupied while the rest of them figured out what was next. Suddenly there was so much to do: sorting through Bobbie's belongings, getting the house ready for David and Jan and their kids. And Tom was entering the final stretch of a—so far—six-year master's degree marathon. He had little time for grief. Although every time *NOVA* came on, his first thought— followed immediately by a pang of loss—was, *I should call Mom.*

Even after the dream, Tom would continue to have moments like that: happy lapses until reality jerked him back to Life after Mom. But something changed for him after the dream, the one he awoke from several weeks after Bobbie's death, the dream that rewrote the circumstances of her death into a script he could more easily live with than the one that had her dying frightened and alone in a stuffy back bedroom. In it, Tom was with his siblings at Ecola Point, the second cape south of Tillamook Head, where Tom had stood many times on the cliff above the ocean and looked south to Haystack Rock and Cape Falcon and a parade of lesser offshore rocks and promontories marching down the rugged coastline in hazy blue silhouette. He and his brothers and sisters were playing on the grass at the top of the bluff above the ocean, chasing one another, carefree as fox kits in the sun, not yet the adults they had become but children still, Bobbie's kids. Then Tom's dream-self looked out to the ocean, to the beach below, and there was his mother: not Bobbie at age sixty-four, with her horn-rimmed glasses, her red Nordic sweater and bob of graying brown hair, but Bobbie as a girl, no older than Tom was himself in the dream, in a somber gray shift of the kind she might have worn growing up in the 1930s. She was walking toward the sea, slowly, as if called. Tom could see her looking back, over her shoulder, at him and his siblings on the bluff, a bemused smile showing her perfect teeth, creasing her cheeks. Tom's perspective then shifted, and he found himself looking, with her, back over her shoulder, to the spruce-draped cliff, the emerald grass on the bluff, the azure sky, and the children—himself and his siblings, playing, laughing. She turned back to the sea, resumed walking, and his perspective shifted once more. Now he was back on the bluff, watching his mother walk purposefully into the ocean until she was gone from sight. Gone, like the last note of a symphony after it fades to stillness: that beautiful, and that final.

Waking alone in his bedroom in Corvallis, Tom felt something shift inside him, like a gate spring pulling closed, the tension released. He knew now that his mother was where she needed to be: called, in his dream, to the ocean, the mother of us all. She'd given Tom the freedom to grow up in the woods and on the water, to take the risks and reap the rewards of a life in the wild, to choose his own path. Bobbie, who had always welcomed him home but never clung. Now it was Tom's turn to let her go. He'd seen what she'd seen, looking over her shoulder from the edge of the sea: that they were okay, would be okay without her. And thanks to her, they mostly were.

15

Mad River Slough

HUMBOLDT COUNTY, CALIFORNIA, 1987

We are like a judge confronted by a defendant who declines to answer, and we must determine the truth from the circumstantial evidence.
—Alfred Wegener, *The Origin of Continents and Oceans*, 4th edition

GARY CARVER SWUNG OFF U.S. 101 at the Trinidad exit and headed up the road toward home. The drive south from Salem, Oregon, had been on a par with the drive north—baby Molly alternately sleeping and crying, toddler Terra bored to the brink of crankiness, he and Deborah frazzled and testy. But the return leg of a family road trip is always a bit worse, coming on top of several days' sleeping in an unfamiliar bed and tending to young children outside of their known environment. Their visit to Oregon had gone well, all things considered; Deborah's mother got to meet four-month-old Molly, and they'd timed the trip to coincide with a Cascadia symposium that Gary had been invited to, at Western Oregon State College in Monmouth, a scant half-hour's drive from Deborah's mother's house in Salem. But everyone was tired now and ready to be home. And, for Gary, there was an added element at play: a fresh salt marsh beckoning.

There had been many meetings similar to this symposium over the past decade, up and down the West Coast, all focused on the same pair of questions: had the Pacific Northwest experienced great earthquakes in the past, and was it likely to in the future? For earth scientists still working out the details of plate tectonics, it was a fascinating conundrum. For the millions of people living on the Pacific Northwest coast and inland, in Vancouver, B.C., and Seattle and Portland and Eureka and towns in between, the implications were huge. Earthquakes had barely figured into the Northwest's building codes, especially in Oregon, the quietest state quake-wise on the West Coast. The prospect of a catastrophic tsunami of the kind Alaskans experienced in 1964 and Chileans in 1960 had not been on virtually any state official's radar. In the Northwest, *tsunami* was widely assumed to mean just one thing: leftover waves from an earthquake far, far away.

Bob Yeats, chair of the geology department at Oregon State University, had called the meeting, prompted by what he'd heard of Peterson's and Darienzo's recent discoveries, along with the work of Atwater up in Washington and Carver in northern California. He had summoned the four of them—they had never before met—along with a spectrum of other Cascadia specialists: believers, unbelievers, and agnostics. Yeats wasn't a Cascadia specialist himself, but he attended the meetings, he read the journals, and he knew the players. He'd heard all the theories: *Cascadia was dead, played out. Cascadia was still a subduction zone, but the oceanic plate was sliding smoothly and aseismically under the continental plate. Cascadia was actually* educting—*going backwards.* If nothing else, it was a good opportunity for all the players to meet one another. And, in the highly competitive world of grant-funded research, to show their hands.

It was John Adams who had first piqued Yeats's interest in the subject in a serious way, back at a geologists' conference in Winnemucca, Nevada, in 1983. Adams, a New Zealander now working for the Geological Society of Canada, had been looking at evidence of turbidites—underwater landslides of the kind thought to be triggered by big earthquakes—in the cores Vern Kulm had sampled on the continental slope off Oregon back in the mid-1960s. "You're going to rock the boat!" Yeats told Adams after hearing him speak. "You're going to scare the hell out of everybody in Portland." But Adams published his findings anyway. Yeats needn't have worried; no one besides geologists had paid any attention.

Adams had come to the symposium, all the way from Ottawa, Ontario. He'd been intending to visit Oregon anyway, to reexamine some of Kulm's ocean cores. Brian Atwater had driven from Seattle with a group from the UW in the seismologists' Suburban, and Mark Darienzo and Curt Peterson had driven up together from Corvallis, a half-hour away. Others had come from California, Colorado, British Columbia. It was, on the surface, much like all the other Cascadia meetings of the past decade: one scientist after another giving his fifteen- or twenty-minute presentation, flashing slides on the screen. But by the end of the day, it was clear to everyone there that something had shifted in the last year, that the believers in an active, quake-producing Cascadia Subduction Zone now outnumbered the non-believers.

Buried marshes? Jerky submergence? For Gary Carver, the meeting in Monmouth had been nothing short of a revelation. While he'd been combing private timberlands and farmers' pastures looking at fractured, folded onshore marine sediments and thinking *earthquake,* these guys in Oregon and Washington had been pulling cores in salt marshes and scraping the mud off coastal riverbanks and coming up with the same conclusion, just coming at it from entirely different angles.

He dropped off Deborah and the girls at the house, threw a few bags out of the car, and backed the car around, promising to return shortly. Deborah didn't seem overly annoyed; after the six-hour drive, she probably needed some space. He got back on U.S. 101, heading south. It took him less than twenty minutes to reach the exit north of Arcata that led to Mad River Slough.

Approaching the bridge, he parked alongside the highway, wet from recent rain, and started walking north and west among the knee-high grasses and sedges. At the edge of the slough, he lay down on his belly atop the springy mattress of pickleweed and saltgrass. He could hear the hiss of cars pouring by on the highway, and the high drone of a single-engine plane overhead, and the rumble of the sawmill at the edge of the bay. As luck would have it, the tide was out. Now Carver reached down the steep slough bank and began digging with his bare hand, clawing at the dark mud with his fingers. The mud, once disturbed, released a sulphurous smell mixing with the tang from the nearby sawmill.

It didn't take but a few minutes. There, under the mud that now blackened his fingernails, he could see the layers lining up below the living peat at the top of the marsh: the shell-flecked sandy mud of the intertidal slough, the band of gritty ocean sand a centimeter or two thick, and below it, a dark layer of old, buried peat. Just the way Atwater and Peterson and Darienzo had described it, the layers spelling out a dramatic backstory below the surface of this quiet marsh. It had taken Carver less than five minutes to find it, digging with his bare hand. Who knew how many more layers, how many episodes of cataclysm, lay beneath the reach of his long arm?

Carver stood and stretched his back. Alone at the edge of the marsh, a few raindrops spattering his bare head and muddy arm, he couldn't help but smile.

16

Stumped

WILLAPA BAY, WASHINGTON, 1987

When a tree is cut down and reveals its naked death-wound to the sun, one can read its whole history in the luminous, inscribed disk of its trunk: in the rings of its years, its scars, all the struggle, all the suffering, all the sickness, all the happiness and prosperity stand truly written, the narrow years and the luxurious years, the attacks withstood, the storms endured.
—Hermann Hesse, from "Trees," *Wandering: Notes and Sketches*
(trans. James Wright)

IT WAS STILL DARK WHEN DAVID YAMAGUCHI was awakened by a voice calling to him from the opposite end of the trailer. "Breakfast!" Brian Atwater exclaimed, far too cheerily for—Yamaguchi groped for his glasses, then his watch—not quite 5:30 Saturday morning. Yamaguchi could smell bacon cooking: apparently Atwater had been up for a while and was already at work in the little aluminum trailer he kept parked at the Bay Center KOA campground, the trailer he used as his field office and kitchen. *What, no breakfast in bed?* Yamaguchi almost quipped, would have had he known Atwater a little better. Instead, he maneuvered an arm up, unzipped his sleeping bag, and reached for his pants.

Yamaguchi had driven straight from work in Vancouver, Washington, the previous evening, a Friday in early May, arriving at the KOA before dark. It was all about the tides, Atwater had explained when he telephoned a week earlier to invite Yamaguchi to join him in the field for the weekend. He'd reiterated the plan when they met up that evening. Atwater wanted to paddle to his study site on the ebbing tide, arriving while the tree stumps were high and dry. That would give them a few hours to work before the tide rose, re-filling the river channel.

Yamaguchi was not one to turn down an invitation for fieldwork, especially work so closely aligned with his particular area of expertise: tree rings in old wood, particularly wood buried and preserved in mud. But today's objective, he knew, was a stretch. Yamaguchi was a dendrochronologist: a tree-ring scientist. Dendrochronology had been launched early in the twentieth century in the American Southwest, where trees are frequently stressed by episodes of drought, a type of stress that forest scientists found was clearly expressed in variations in the trees' annual growth rings. These rings result from the annual addition of light-colored "earlywood"—bulbous, thin-walled cells— during a tree's summer growing season interspersed with thinner layers of darker, thicker-walled, small-celled "latewood." Dendrochronology was still a young field; only a decade earlier a University of Washington forester had managed to successfully apply these techniques to Douglas firs in Washington's Cascades. Until then, no one knew if Pacific Northwest forests would be too well watered, too "complacent," to show measurable stress the way trees in arid landscapes did. Atwater wondered if these trees—now stumps—all died at the same time. If so, the cause might have been saltwater inundation of their roots following a big earthquake: the coastline subsides, seawater flows in and doesn't flow back out, and the shoreline trees die. He was hoping Yamaguchi could work his dendrochnology magic in the temperate rain forests of the Washington coast, where the annual rainfall can exceed five or six *feet*, almost half again as much as what falls on the forests of the western Cascades. Talk about complacency: these trees were as well watered as any on the continent. Would trees like that even show discernible deviation in their rings from year to year?

Those doubts multiplied when Yamaguchi, paddling in the bow of the Grumman, got his first glimpse of the stumps: they were roots mostly, thin and gnarly, black and half-rotted. Tree roots put on annual growth rings, just as tree trunks do, but they're harder to read. Unlike the typical columnar tree trunk with its concentric growth rings, tree roots twist and turn, balloon and shrivel, in their effort to locate water underground, in the soil and sand, among rocks and other roots. To have half a chance of getting meaningful tree-ring data, Yamaguchi would need trunk wood. But the trees these roots had once supported and fed had long since rotted away. All that remained of the trunks was a few inches, a foot at best, of stump preserved along with the roots in the now-eroding bank of the Niawiakum River.

Yamaguchi, clad in hip waders, stepped out of the canoe and helped Atwater haul the boat farther up the bank. The first challenge, he found, was managing to stay upright on the slippery stream bank. Wrangling a chainsaw while standing on mud was the next; it was tricky enough in the rock and ash landscape at Mount St. Helens, Yamaguchi's usual venue. Undaunted, he strapped his Kevlar chaps over his rubber waders, fueled the saw, half-walked and half-slid to the most promising-looking stump, and yanked on the starter rope.

He did what he could, expertly incising the stump until he had a wedge some fifteen inches long extending from the outermost wood to near the center of the heartwood, freeing the dead tree's musky fragrance to mingle with the rotten-egg odor of disturbed mud and the saw's own gassy fumes. The results were, as Yamaguchi had feared, disappointing. The long-dead tree's growth rings were so wide that the trunk sample he'd extracted captured only twelve or fifteen years' growth. At Mount St. Helens, where he had done his doctoral research and was still working on a post-doc, he had developed a rule of thumb: it takes at least two hundred rings—two hundred years' growth—to confidently compare one tree's life history with another's and, sometimes, identify the season and year of a dead tree's demise.

He held the wedge of wood out to Atwater, grimacing an apology; he simply wouldn't be able to do much for the geologist. With no needles, no bark, no cones to go by, Yamaguchi wasn't even certain what species of tree he was dealing with.

He cut chunks from a couple more stumps and cross-sections of some protruding roots before powering off the saw. "Brian, I can't help you," Yamaguchi said. "I'm not getting enough rings to work with."

"Cut into those!" was Atwater's response, pointing at more stumps scattered atop the salt marsh. So Yamaguchi clawed his way up the bank and walked first to one stump, then another, still frowning. At least these stumps weren't buried in mud, but they were too badly decayed to be of any use. Yamaguchi shook his head and walked back to the riverbank.

"It's just too messy," he said with genuine regret. "I'm not going to be able to help you." What he really needed, Yamaguchi explained to Atwater, was intact trunk wood. Not decomposed roots, not the eroded remnants of stumps rotted down to the level of their roots, but the tree itself, aboveground, enough to get a clean cut from heartwood to bark, or at least to outermost wood. It any of these trees had a story to tell—a story of death by earthquake—Yamaguchi would need the detail provided by the outermost rings. Was it a slow decline, or did it succumb suddenly? Had the tree been struck down in youth, or well into a long and full life?

It had been worth a try. The two men had met six months earlier, when Atwater gave a brown-bag talk to staff at the Cascades Volcano Observatory in Vancouver, where Yamaguchi was two months into a year-long post-doc position, having just completed his doctorate at the University of Washington. The story Atwater told to the crowd of mostly volcanologists that noon hour—surprising evidence he'd recently discovered in the marshes surrounding Willapa Bay of successive great earthquakes and tsunamis on the Washington coast over thousands of years—made perfect sense to this audience of volcano cowboys: geologists drawn to the high-risk world of monitoring and studying active and potentially active volcanoes around the world. Cascadia Subduction Zone? They hadn't necessarily heard it called by that name, but they were all aware of its existence. A *seismically active* subduction zone? It made sense. There had to be some mechanism feeding magma to the Cascade volcanoes, one of which—Mount St. Helens—had demonstrated its vigor just six and a half years earlier. Atwater clicked through the slides in the carousel projector, flashing diagrams of marsh stratigraphy on the Niawiakum River, photographs of the riverbank with its black and brown layers of mud and peat and occasionally sand and the gnarled appendages of roots from long-dead trees protruding from the mud.

Geology with wood sticking out of it, Yamaguchi recognized in a flash: *that* he understood. It was the focus of his post-doc at the CVO: dating the rings of trees killed and buried by old lahars—volcanic mudflows— on Mount St. Helens, trees exposed by flooding following the most recent eruption in 1980. At Atwater's last slide, Yamaguchi couldn't help but raise his hand.

"Hey, I know a young guy who could probably help you," he piped up, smiling winningly. "He's going to have a lot of time on his hands very shortly!" At that, the room erupted with laughter. They all knew who that young guy was.

Now it looked as though Yamaguchi would have to look elsewhere for a job when his post-doc ran out in the fall. From the looks of it, tree-ring science was not going to provide the magic bullet the geologist was seeking: the key to fixing the date of the last great Cascadia earthquake.

Two months later, Yamaguchi was surprised to get another phone call from Atwater. He recognized the voice right away—that deliberate, measured speaking style—but this day the geologist's words were edged with an unmistakable tone of excitement.

"Dave, you have to get back here," Atwater told him. "I think I've found what you were looking for."

TREE TRUNKS, NOT ROTTING ROOTS but standing snags: that's what Yamaguchi had said he needed, and that's what Atwater had found. As his second field season in southwestern Washington got rolling, Atwater had begun exploring more of the region's coastal waterways, canoeing farther up coastal rivers he'd only glanced at the previous year before focusing his efforts on the Niawiakum.

It was on a deeper foray up the Copalis River, north of Grays Harbor and halfway up the Washington coast, that he had stumbled upon a grove of silvery weathered snags, some tall and some short, all clearly long-dead trees standing in a high marsh about one and a half miles from the river's mouth at the community of Copalis Beach. His first thought upon spying the trees: *Call Yamaguchi*. His second thought: *Something besides subsidence must have killed these trees*. If, as he believed, sudden land-lowering during an earthquake was the cause of death of the Niawiakum trees, it didn't make sense that these trees would still be standing and the trees along the Niawiakum would all have rotted to the ground. Not if they had died at the same time, in the same event. Assuming, of course, that they were the same species of tree.

Which, as it turned out, they weren't.

That much was obvious the moment Yamaguchi's chainsaw bit into the base of the first snag in what Atwater was calling the Ghost Forest. Both men smelled it: that resiny tang, the unmistakable perfume of cedar. These standing snags were western redcedar, an iconically long-lived and rot-resistant tree, thanks to natural fungicides in its heartwood. Redcedar is the tree Native people on this coast had long built canoes from, what they and others used to shingle roofs and houses. The roots sticking out of the banks of the Niawiakum had smelled merely musty when cut. Turns out they were Sitka spruce, another iconic coastal tree but one distinct from redcedar in key respects. Sitka spruce grow only within a few miles of the coast; they grow quickly and can live for centuries but typically not, as redcedar can, for a millennium. And rot is actually one of the spruce's talents, part of its genius, its reproductive strategy: new spruce seedlings take root almost exclusively in the rich, rotting remains of spruce trunks, snags, and stumps.

Within minutes of his arrival at the Ghost Forest, Yamaguchi had extracted a chunk of wood from one of the snags and was beaming at the narrow rings of trunk wood: more than one hundred rings, maybe two hundred. He turned to Atwater, proffering the triangular slab, grinning. "I can work with this," he said.

"So how old do you think these trees are?" Atwater asked eagerly.

"Well, if we were on Mount St. Helens, these snags would have been standing here less than five hundred years," Yamaguchi ventured, taking the measure of the trees' decrepitude, "but how much less, I don't know. That's my best guess. This is a different situation: damper environment, different decay rates, different tree species. If I had to guess, I'd say less than five hundred years."

The trees had been alive for maybe two hundred years, from the tree's ring count. Dead for five hundred or less. As old as Dante, or Chaucer? Growing here, maybe, while Gutenberg was building his first printing press? Already dead by the time da Vinci painted *The Last Supper*? Before Shakespeare first staged *Macbeth*? Dead by 1792, when Captain Robert Gray sailed the *Columbia Rediviva* into the mouth of what he would name the Columbia River? Definitely dead before Lewis and Clark first set foot on this coast in 1805, else the mountain men, merchants, settlers, and government agents who followed close on their heels would have reported the calamity, left mention in a journal, written home about it, something.

Nowhere in the historical record was there mention of a very large earthquake off the Pacific Northwest coast, nor of the tsunami that would have followed it. But that record is less than two hundred years old. A smattering of European mariners may have made landfall or wrecked on the Pacific Northwest coast in the 1600s or early 1700s, but none had stayed, or if they did, they didn't keep notes. Not until 1769 was the first Spanish mission established in southern California. The coast between Alaska and southern Oregon's Cape Blanco wasn't even charted until the 1770s, and not until the late 1780s did trade commence between American and European mariners and Native people of that region. It would be a couple of decades more before anyone with the means and motivation to keep a written record—a diary, an inventory, weather observations, letters to mother, anything—would arrive and stay put on these shores. Who knew, Atwater wondered, what kinds of cataclysms might have been visited upon this coast before then?

Yamaguchi, meanwhile, had retaken possession of the wedge and was staring at it, taking its measure. Maybe two hundred rings, with clear width variation ring-to-ring, year-to-year. Here and there were pockets of rotten wood, the bane of dendrochronologists working, as they typically do, with pencil-thin cores carefully extracted from living trees. But these trees had been dead, from the looks of it, for centuries; no need for delicacy with them. Yamaguchi knew how to sample these long-dead trees; he'd been doing the same thing to the lahar-killed trees at Mount St. Helens all that spring and summer. Insert the saw's 24-inch bar horizontally, reinsert it at a slight angle still aimed at the tree's heart, do it again, and eventually you will pull out a wedge of wood that, if you have aimed the saw right, catches every ring, from every year the tree was alive.

Now Yamaguchi had a way to contribute to the Cascadia story, or so he hoped. His field time that summer was mostly spoken for—he had tons of work still to do on Mount St. Helens. But he managed to shoehorn in a couple of weeks on the coast, paddling rivers with Atwater or with his own student assistant in the red plastic Coleman canoe his dad had won in a grocery store contest. A canoe was the vehicle of choice for a coastal forester as well as a coastal geologist: the trees Yamaguchi sought were all near rivers, there were no roads to these remote groves, and the barge-like canoe easily handled the volume and weight of saw, fuel, and piles of five-pound cedar slabs.

Sitka spruce don't like saltwater, but they are more tolerant of a little brackish water than are cedar and thus tend to be found at lower elevations, closer to the waterline. That fact was key to the story that began to suggest itself to Atwater and Yamaguchi as they continued their explorations: a story of one tectonic plate lurching forward in a cataclysmic quake and, in that process, resetting the land on the southwest Washington coast at an elevation three or four feet lower than it had been a moment earlier. That subsidence turned marshes into bays and low-elevation forests into marshes, killing the cedars and many, but not all, of the spruces growing there. Then, over decades and ultimately centuries, the land began its slow rise. The dead spruces rotted and fell and their trunks decomposed, though the cold tidal mud preserved their stumps and roots. But the dead cedars stayed standing, their bark and outermost wood worn away by the storms of hundreds of winters. Around those snags and tall stumps—some of the dead cedars had been salvage-logged—a new forest of saltwater-tolerant spruces had grown up.

Yamaguchi learned how to spot the spiky tips of cedar snags poking above the green tops of the spruces. He would land the canoe, unload his gear, and fight his way through the dense underbrush, sometimes using the chainsaw to cut tunnels through the salal and huckleberry to reach the base of the tree. Once there, he would fire up the saw and extract a neat wedge of wood, taking meticulous notes about its location. Before departing, he would nail an aluminum ID tag the size of a quarter to the tree. Yamaguchi had been using pre-numbered tags to mark sampled trees since he'd started his tree work at Mount St. Helens, where he had tagged 576 stumps for his dissertation research. He had tagged more than a hundred lahar-buried trees after that. By the time he began collecting tree-ring data on the Washington coast, Yamaguchi's tags numbered in the 700s.

When did these trees die? The dendrochronologist had a growing collection of slabs from trees with a long and clear ring history. But the trees' bark and outermost wood had long since weathered away—exactly how many rings were gone was impossible to tell. Without those outermost rings, he had no way to know exactly how old the trees had been at their death, or even to be sure they had died suddenly, as he supposed they would have in an earthquake, rather than wasting away over a period of years.

Even if the snags' outermost rings had been intact, all that would reveal was the age of the tree when it died—not *when* it died. Yamaguchi needed a comparative tree-ring record from living trees, or trees felled on a known date. Then he could compare the patterns of tree rings—the wide and the narrow rings, the fat years and the lean years, the stressful periods and the periods of easy living—between the known "index" trees and the mystery trees. Line them up, one against the other, matching the ring patterns, and you could be pretty confident what happened when. At least, that was the theory. It had worked for trees in the American Southwest and old-growth Douglas firs in the Cascades, and Yamaguchi had extended that work to lahar-killed Doug firs on Mount St. Helens. Brian Atwater was hoping that David Yamaguchi could make it work with western redcedars in the temperate rainforest of southwest Washington. All Yamaguchi needed were some living trees of the same species, growing in the same or similar conditions, trees that he could core, or stumps of same-age trees felled on a known date.

But there were no such trees, none that he knew of. Beginning in the late 1800s, the Washington coast had been thoroughly logged, virtually every tree of any size taken. There were plenty of trees on the coast, lush forests full of them, but they were all second- or third-growth: young trees planted after all the big, old trees had been felled and hauled to the mill. Missing were old, living trees—"witnesses," as Atwater began calling them—that Yamaguchi could use to calibrate the death date of the "victims."

Then serendipity intervened in the form of a log truck.

It was no ordinary load of logs that caught Atwater's eye late that summer of 1987. Typically one log truck in southwestern Washington can haul twenty, fifty, even one hundred skinny logs to the sawmill or pulp mill in one load. But resting on the trailer of this log truck was but one trailer-length chunk of redcedar nearly eight feet in diameter.

So not all the big, old trees were gone, Atwater realized, watching the truck disappear around a curve, headed perhaps to a sawmill in Aberdeen or Raymond. Someone, somewhere, was logging old growth.

Atwater headed south on U.S. 101, the direction the truck had come from, past the mouth of the Niawiakum and the road to Bay Center, to where the southern end of six-mile-long Long Island—a low, forested finger of land in the middle of Willapa Bay—nearly touched the mainland. There he noticed a truck ferry, one he must have driven past dozens of times: a barge designed to transport log trucks across the narrow channel between the island and the mainland landing. Even now, as Atwater sat in his pickup truck idling on the shoulder of the two-lane highway, another truck emerged from the forest at the edge of the island and rumbled slowly to the ferry landing, loaded with another venerable chunk of cedar.

Yamaguchi needs to see this.

Long Island, they learned, still harbored groves of ancient cedar trees. Four years earlier, the U.S. Fish and Wildlife Service, which managed the island as part of Willapa National Wildlife Refuge, had acquired all 1,621 acres the Weyerhaeuser company owned on the island, including cedars up to a thousand years old. In exchange, Weyerhaeuser received cutting rights to several hundred acres of second-growth timber elsewhere on the island. According to the deal, if the value of the second-growth timber was too low to compensate for the value of the old-growth cedar it had relinquished—which, it turned out, it was—Weyerhaeuser could continue to log old growth. Environmental activists had since convinced Congress to give the company more than $6 million to preserve the island's most treasured ancient cedar grove. Meanwhile Weyerhaeuser was continuing to cut and haul out not just younger second-growth trees but some of the last ancient western redcedars still standing on the southern Washington coast.

Yamaguchi and Atwater's assistant, Ken Bevis, waited until mid-afternoon, after the loggers had quit work for the day, before making a reconnaissance trip. They didn't want to be anywhere near an active logging site while trees were dropping, nor tangle with log trucks on the island's narrow roads, nor be asked about what they were doing there. Weyerhaeuser wouldn't miss the wedges Yamaguchi hoped to cut from the stumps, and it was easier to avoid questions.

They paddled across the narrow channel and landed the canoe on the tide flat next to the ferry landing. Then they shouldered their gear—chainsaw, gas, oil, safety goggles, hard hats, gloves, chaps—and took off up the gravel road curving along the shore. Robins darted among the alder trees edging the road, which curved and rose, heading north into the heart of the island.

The logging operation was on a rise less than a mile up the road. You couldn't miss it: a messy tangle of broken limbs and mud, with sunlight streaming in through holes punched in the forest canopy, illuminating a forest floor that, for millennia, had seen only passing shafts. And huge stumps. Walking from the new clearing and into the not-yet logged portion of the grove, it felt to Yamaguchi like the air temperature dropped twenty degrees. Slender deer fern licked his boots, and salal and red huckleberry grew taller than his head, stretching toward the slivers of sky visible through the lacework of the huge trees' gnarled canopies hundreds of feet overhead. An hour or two earlier, this place would have been a cacophony of destruction: buzzing chainsaws, growling bulldozers excavating a stump-free forest bed to receive valuable falling cedars, the thunderous thud and shattering crash of giant trees striking the earth, the snap-crackling of skidders moving the downed trees to the roadside, the rumble of idling diesel log trucks. Now all was silent but for the low whine of insects.

It was Mount St. Helens all over again for Yamaguchi: a field of venerable stumps, devastating for a lover of living trees such as Yamaguchi to witness, exhilarating to Yamaguchi the dendrochronologist. Mount St. Helens' 1980 eruption and the swath of forest destruction it cut through the surrounding forest had been what prompted Yamaguchi to take up tree-ring science to begin with. The force of the eruption had torn trees close to the crater from their bases, scattering them like pick-up sticks. Farther out, trees had simply blown down, all of them lying in the same orientation, like an animal's flattened fur. Farther yet was a forest of standing dead Douglas firs, singed brown by the blast's hot gasses. Those standing snags became the focus of Yamaguchi's PhD research, examining the five-hundred-plus-year calendar embedded in the forest's ancient Douglas fir trees to learn how forests reestablish themselves and thrive after volcanic cataclysm. Timber companies had negotiated a controversial deal with the U.S. Forest Service to "salvage" many of these trees, cutting them off at the base and hauling them to market. There had been important ecological reasons to leave the dead trees in place, serving the functions that all dead trees serve: habitat for wildlife, growth medium for other forest plants. Dead standing trees are part of the recipe for forest recovery and natural succession. Yamaguchi would have preferred to see the forest around the volcano recover on its own, complete with snags. But the timber companies, seeing dollar signs in singed but otherwise intact old-growth Douglas firs, had argued that standing snags would invite insect infestation and wildfire, threatening the adjacent living forest. Soon, an army of loggers had begun felling and hauling away the blackened trees. But each felled tree left behind a smooth stump revealing hundreds of concentric rings, like a bull's eye. Yamaguchi, who until then had been drilling skinny tree cores and storing them in soda straws, suddenly saw in those stumps a golden opportunity: more, larger, and better rings served on a silver platter. He had put away his coring device and began using a chainsaw to extract wide wedges of wood from the stumps—tree-ring records far more telling than a core of wood the width of a pencil. Ultimately he would sample nearly five hundred Douglas fir stumps. And it paid off. By comparing the ring patterns of trees in different parts of the forest around the volcano, Yamaguchi not only assembled a nuanced picture of forest succession and ecology post-eruption, he corrected earlier estimates of the date of the volcano's previous eruption (A.D. 1800, not 1802) and provided new estimates for the two that preceded it (the dormant seasons of 1479–80 and 1481–82).

Timing: once again, Yamaguchi's was impeccable. Just when he'd turned his attention to Douglas fir dendrochronology six years earlier, salvage logging had provided him with the raw material to unravel one volcano's geologic history. Now, just as he had despaired of finding contemporary corollaries of the tsunami-killed snags, Weyerhaeuser had handed him a field of fresh stumps. These old cedars—among the last of their kind on this coast—needn't have died to be useful to him; he could have cored the living trees. But had loggers not started logging, he wouldn't have known the trees existed. And as long as they were logging, the stumps they left provided him with the best possible reference material.

Yamaguchi didn't pause long before getting to work. He and Bevis spent what was left of that afternoon and all the next working, Yamaguchi sawing and Bevis carrying the cedar wedges back to the canoe.

Not until November—after snow had blanketed the high country at Mount St. Helens, shutting down his fieldwork there—was Yamaguchi able to turn his attention to the pile of Long Island cedar wedges stacked in his living room. He was still living in Vancouver, having negotiated a six-month extension of his post-doc. The day before Thanksgiving, he loaded the slabs into the back of his station wagon and headed to Seattle. He spent the next day with his parents, enjoying a traditional Yamaguchi family Thanksgiving dinner—turkey, rice, pumpkin pie. He spent the rest of the long weekend at the university, taking the measure of the Long Island stumps and the Ghost Forest snags.

His work began with sanding. In dendrochronology, surface is everything, and the UW School of Forestry had a killer woodshop. It might take 600 grit or even 1200 to get a clear reading of a piñon pine's narrow rings. But Douglas fir slabs revealed their history with just 400 grit—what you might use to take the paint off a car's quarter panel prior to priming—and the same, Yamaguchi found, was true for western redcedar.

Even before he had all the samples sanded and logged, Yamaguchi could see he had something. Nineteen of the samples Yamaguchi and Bevis had cut on Long Island had tree-ring patterns virtually identical to one another. These were old trees, with six hundred, eight hundred, up to nearly one thousand rings. Alive since at least the fourteenth century, some even since the tenth, and still thriving until last summer, these trees almost certainly overlapped the lifespans of the Ghost Forest cedars. The Long Island grove stood about 160 feet above sea level—low enough to chronicle the same kind of reactions to changes in their tree rings as those in the Ghost Forest through hundreds of years of climate fluctuations, but high enough to have escaped saltwater inundation when the coast suddenly subsided during the last great earthquake. In short, witnesses.

Now Yamaguchi had a calendar for estimating when long-dead cedar snags on the Washington coast had died—at the Copalis Ghost Forest and elsewhere. But first he had to find out whether other cedars in other locations along the coast had died at the same time. That would be the test for death by earthquake subsidence rather than from some phenomenon particular to a single tree or a grove of trees. And that would be his next assignment: finding more long-dead cedar snags on more rivers on the Washington coast.

If he found that cedars in distant locations along the coast did die simultaneously, certainly the next question would be *when*? Lining up the ring-width patterns of the Long Island and Ghost Forest trees, Yamaguchi could see that the Ghost Forest snags had been alive until at least 1680. But how much longer? They had been dead so long, their bark and even some of their wood had been weathered away. To get a reliable death date, he would need to find an earthquake-slain cedar snag intact nearly to the bark, at least to its outermost wood.

But Yamaguchi had reason to believe that these cedars hadn't died much past 1680. His source was a most reliable informant: bark beetles. They infest dying or dead trees, eating and reproducing and burrowing tunnels perhaps a quarter inch wide in the *phloem*—the food-conducting tissue between the outer bark and the trunk wood. Yamaguchi could still see traces of beetle "galleries" engraved on the outside of some of the Ghost Forest trees. When remnants of beetle galleries were discernible on a snag, Yamaguchi could be fairly certain that not more than a decade or two worth of outermost wood was missing from that tree.

Bark beetles never lie.

17

Ataspaca Prospect

TOM, 1988

TOM HORNING SHUT OFF THE SLIDE PROJECTOR and sat down at the stained and dusty table at the front of the geology lab. He glanced over his shoulder, cringed slightly at the clock over the door: he'd run almost twenty minutes over the half hour scheduled for his thesis defense. *Par for the course.* A copy of his master's thesis sat on the table next to him: 402 pages, not counting the maps he'd stayed up late coloring the night before. "The Geology, Igneous Petrology, and Mineral Deposits of the Ataspaca Mining District, Department of Tacna, Peru." Five pounds of paper, 1¾ inches thick. Seven years of his life, collated and comb-bound with a black plastic spine.

In front of him, across the table, sat his committee of three faculty members. Beyond them, crowded around the room's other two long lab tables, was his cheering section: twenty or so friends plus his sister Judy, down from Portland. Behind them, at the back of the room, ran one long, horizontal window that seemed to suck light rather than cast it into the daylight basement lab that dreary January afternoon. All Tom could see was a dark slope of muddy ground and the round, gray grid of a manhole cover. He wished he could have spent a couple more days refining his talk. But he'd already delayed his thesis defense several times. No one knew its shortcomings better than Tom himself. But sitting there, looking at the faces of his committee members, he granted himself a moment of satisfaction. It was clear from their expressions that he had blown them away.

The Ataspaca prospect, where he'd done his research, sat at nearly 13,000 feet elevation on the west flank of the Cordillera Occidental in Peru's southern Andes Mountains, at the edge of the Atacama Desert, the driest desert in the world. Every day of the thirty-four days he spent at Ataspaca, Tom and a fellow grad student from Oregon State would leave their boarding house twelve miles outside the town of Tacna and drive an hour and a half up into the mountains, hugging the cliffs and rounding the hairpin turns above steep ravines cradling the glittering remains of vehicles that had failed to negotiate those curves. Hardly anything—plants, animals, people—lived on those high, arid hillsides. Nearing Ataspaca, the road dropped into a canyon where snowmelt from the glaciated volcanoes above allowed subsistence farmers to eke out a living raising alpacas in small adobe corrals and growing beans in terraced, irrigated fields. The study site itself was rugged and rocky and cut by deep canyons. What vegetation there was on the prospect was limited mostly to cacti, hardy grasses, and a few low shrubs able to survive without much warmth or water. The sun shone nearly every day, though the wind, blowing off the fields of snow and ice at the summit of the Cordillera just above Ataspaca, could be biting. Tom spent his days mostly hunched over, scrutinizing every inch of his two-square-kilometer study site. Around noon he would put down his hammer and pick and straighten up, stretching his back. Then he'd settle himself on one rock pile or another and unwrap the sandwich the boarding house had packed for him—a hot dog, sliced lengthwise, between two pieces of brown bread smeared with a thin sheen of mustard—and eat it while gazing at distant snow-capped volcanoes framed by a startling blue sky. If Tom didn't eat his sandwich within two or three minutes of unwrapping the waxed paper, the bread would be dry as a cracker. By the end of that summer, Tom had felt as desiccated as his sandwich.

Tom had been aiming at a master's degree in economic geology—the geology of ore bodies that can be extracted from the earth for profit. In short, prospecting. The state-owned mining company Centromin Peru had been digging exploration tunnels at Ataspaca and had started work on an office-and-lodging facility—tangible evidence of Centromin's high hopes for the site. Geologists had already recognized that the mineralization at the Ataspaca prospect was hosted by *skarns*: metamorphic rocks produced when hot water deep in the earth facilitates an exchange of calcium and other elements between rocks in different parts of the ore system, creating new minerals such as red garnet and green pyroxene. Such skarns may host a wide variety of exploitable minerals: iron, copper, zinc, lead, gold. Tom's aim was to identify and map the minerals—the veins and arteries of the underground ore bodies at Ataspaca—to better understand this site's geologic history and its potential for exploitation as well as that of similar sites in other parts of the world.

The work he'd done in Peru was groundbreaking, his professor had told him—worthy of publication in the *Journal of Economic Geology*. But that hadn't worked out. Along came a couple of guys from Stanford University investigating an ore body in southwestern Nevada with a configuration similar to Tom's in Peru. They got their paper in six months ahead of Tom, stealing his thunder. They had grants to pay for their atomic absorption spectrometry and arc emission spectrography; Tom had had to stop work on his thesis and paint houses to cover the costs of his lab work.

By the time Tom had settled on geology as his undergraduate major in the mid-1970s, there was hardly a geologist in the world who wasn't on board with plate tectonics; it had become the field's lingua franca. No one doubted that the Nazca Plate was subducting under the South American Plate; frequent earthquakes in Chile and Peru attested to the plates' motility. The basic geology in Peru mirrored that of the Pacific Northwest: an oceanic plate butting heads with a continental plate, with a line of volcanoes—the Andes in Peru, the Cascades up north—just inland. But *was* the Juan de Fuca Plate off the Pacific Northwest coast subducting under the North American Plate? With few earthquakes recorded in the region, it was hard to say exactly *what* was happening. The plate boundary there had remained a mystery throughout most of Tom's graduate school years, one he hadn't focused on, though more than once he had overheard conversations in the corridors of Wilkinson Hall about a geologist in Seattle who had found evidence of great earthquakes and tsunamis in the marshes of southwest Washington. Curt Peterson, on the research faculty at OSU, had apparently made similar discoveries at the same time on the Oregon coast. Tom liked to think he had planted that seed of interest in Curt years earlier, back when both were grad students at OSU and they had sat together on Tom's lawn in Seaside one Saturday, shucking razor clams, and Tom had regaled Curt with tales of his own close brush with death-by-tsunami.

But Tom had paid scant attention to the Cascadia chatter. He was nearing the statute of limitations on completion of master's theses, and if it wasn't about skarns, he didn't have time for it. That final year there were many days when Tom found it hard to focus on geology at all.

Bobbie had taken such a keen interest in Tom's research, had been his unwavering, one-woman cheering squad, commiserating with him when a pair of researchers beat him to the punch. "You're like Alfred Russel Wallace, aren't you?" she once remarked, conjuring the nineteenth-century naturalist whose theory of evolution due to natural selection was ultimately eclipsed by the writings of the wealthier, better-connected Charles Darwin.

A bit of hyperbole, granted, but Tom liked hearing it just the same. She had been so eagerly looking forward to watching Tom's thesis defense. She had missed it by one year, almost to the day.

A committee member's question pulled Tom back into the present. "Tom, about the origin of the metals in the ore here," Vern Kulm began. But instead of putting him on the spot, the oceanographer pitched Tom a softball, nothing he couldn't handle. Tom had expected some grilling, some attempts to catch him off-guard—the usual academic bullshit. But it was hard for them to get too picky with a master's degree candidate sitting next to five pounds of PhD-caliber material. He navigated their question-and-answer period, sat quietly in the corridor outside while his committee held its obligatory conference, and accepted their handshakes and his friends' high-fives after he was called back in. Then he picked up his thesis and, hugging it to his chest with one arm, ascended the stairs and headed out the front door of Wilkinson Hall, walking off the campus where he'd spent much of the second half of his life so far and into whatever was next.

18

Thirty Minutes

CANNON BEACH, OREGON, 1993

*Had the pilgrims landed at Haystack Rock near Cannon Beach
instead of Plymouth Rock, the historic memory of this last
[seismic] event would have no doubt been more strikingly instilled
in our consciousness.*
—James Bela, from "The Dilemma of Great Subduction, or How I
Learned to Love Politics and Stop Worrying About the Earthquake"

ALFRED AYA, STANFORD UNIVERSITY CLASS OF 1950 medieval philosophy
major, a senior statistician with PacTel in San Francisco, knew he'd found
just the place to retire when, on a driving trip up the Oregon coast in the
winter of 1981, he got stormed in for a week in a motel in Cannon Beach.
The town was small and attractive. It was the off-season and dead quiet; half
the shops and restaurants downtown were closed. The Bell telephone system
where he had worked for twenty-five years was on the verge of breaking up;
it was time to think about collecting his pension and getting out. And he
was tired of the Bay Area's traffic, the congestion, the crime, the nagging
worry that the next earthquake might be the Big One. He already knew a
couple who had retired to Cannon Beach; they urged him to join them. To
read, to write, to live quietly, to walk on the beach and philosophize, and
finally to die: that, in a nutshell, was the pragmatic Aya's plan for what he
referred to, just a little sarcastically, as his "golden years."

Joining the board of the Cannon Beach Rural Fire Protection District
was never part of the plan, and yet there he was at his first meeting in May
1985, just months after moving to town. "I don't know anything about
firefighting," he'd objected when a friend had urged Aya to fill the vacancy
on the board created by that same friend's departure from it. "You'll learn,"
the friend told him, stressing the minimal time commitment required:
"About an hour a month."

So Aya was entirely unprepared for the topic of discussion that arose at his first meeting, as the board reviewed the budget for the upcoming fiscal year. "What's this $1,000?" asked another new board member sitting two chairs away, pointing a fat finger at a line item on the page before him. The man—owner of a trucking company—was built like a bear, his hands the size of hams, or so it seemed to Aya.

That, explained Fire Chief Gary Moon, was the amount Chief Moon proposed adding to the district's tidal wave alarm system fund in the next fiscal year. One year earlier, the board had voted to set aside $11,000 toward purchase of such a system, Moon said; the plan had been to add to it in subsequent years until the district could afford to buy and install such a system.

"When did you last have a tidal wave here?" Al's fellow board member inquired. Chief Moon had to think for a minute.

"1964," he finally answered, pausing a moment longer to do some mental math. "Twenty-one years ago."

"That sounds like a pretty rare event," the man said. "How do you know when one is coming?"

"They tell us."

"Who's *they?*"

"The government."

"Which government?"

"The sheriff."

"Who tells the sheriff?"

Chief Moon didn't have a ready answer.

"How does anyone know a tidal wave is coming?" the man pressed further.

"I suppose it's when the ocean crosses the high tide line," Moon ventured, unnerved by now.

"I don't need someone to tell me the water's rising!" the ham-fisted man exploded. He moved to dissolve the fund. Someone else seconded it. Aya voted with the majority to dissolve the fund and use the reserved $11,000 to meet more pressing needs.

And then a couple of days later, on a beach walk, Aya stumbled across two youngsters building a sand castle. They couldn't have been more than six and eight years old. They were by themselves, as far as Aya could tell.

It bothered him, seeing kids all alone on the beach like that, no adult with them. His own mother would never have allowed it. That had been an ironclad rule on beach vacations when Aya himself was growing up. That, and *never turn your back on the ocean*. Once, as a child, he had been caught by a sneaker wave, had been pulled off his feet and tumbled in the surf before he was able to thrash his way out. The experience had left an indelible impression on him, a wariness and respect for the power of the ocean.

Are those children on the beach more vulnerable than people realize? Aya wondered. Are we all? Had he and the rest of the Cannon Beach Rural Fire Protection District board made the right decision?

The next day he headed to the public library in Seaside to research tidal waves (*tsunami* was the preferred term, he learned), combing through back issues of magazines: *Science, Nature, Smithsonian, National Geographic*. He vaguely remembered hearing about an earthquake in Alaska back in 1964; he *hadn't* known that it had generated a tsunami that had traveled all the way to Cannon Beach, Oregon, taking out the bridge at the north end of town, knocking out electricity, and flooding the playground at the elementary school. Nor did he remember reading anything about an earthquake in Chile just four years before that; at magnitude 9.5, it had been the most powerful quake ever recorded on Earth, he learned. That quake had generated a tsunami that had not only inundated Chile's own coast but had killed sixty-one people in Hilo, Hawaii, more than six thousand miles away, and 138 people in Japan, four thousand miles farther. And there was the tsunami from the Aleutian Islands in 1946 that killed at least 170 people in Hawaii, and one from Peru in 1868 that washed away houses and bridges as far away as Hawaii and New Zealand. The statistics, the human costs they implied, were staggering, as was the frequency of tsunamis in the Pacific, which seemed to be a regular tsunami factory.

Aya had had no idea.

Aya met Moon at the fire station downtown and heard from the chief how he had been blindsided, dumbfounded in fact, by the sudden turn in the conversation at the previous board meeting. A couple of years earlier, Moon now explained to Aya, he had come back from a meeting with county emergency management officials fired up about tsunamis and tsunami evacuation. Since the district's volunteer firefighters now relied on radios to be summoned, Moon figured the siren on top of the fire station could be repurposed strictly as a community tsunami alarm. But a year earlier Moon had discovered that the old siren, long disused, was corroded and no long operable: hence the tsunami warning-system fund. There were more tsunami advisories for the Oregon coast than people in Cannon Beach realized, Moon disclosed. Not infrequently did he receive notifications from the tsunami warning center in Palmer, Alaska, which issued warnings for Pacific coastal locations stretching from the western end of the Aleutian Islands down through California, about earthquakes at one place or another that might produce tsunamis on the West Coast. He didn't talk about them because he didn't want to frighten people. To date, in his administration as chief, none of the alerts had become full-fledged warnings.

At the next monthly meeting of the Cannon Beach fire district board, Aya shared his research and proposed that the board reinstate the $11,000 tsunami warning-system fund. The board agreed and added the additional $1,000. And Aya was appointed to the task of researching and recommending such a system to the board.

In 1985, no community on the West Coast had in place a warning system specifically for tsunamis. What Aya had in mind was a siren that would not only sound a loud wail but that had voice capability, allowing emergency managers to provide instructions to tourists and locals about leaving the beach and other low-lying areas and heading uphill. The only comparable systems Aya could find in use in the United States were in communities with nuclear power plants. Those became his model. In December 1986, four electronic, voice-capable sirens were delivered to the Cannon Beach fire station.

Then, just four months later, came the meeting that changed everything for Aya.

It was billed as a routine meeting organized by the Oregon Office of Emergency Management and held in Coos Bay. Not until it got under way did Aya learn that an additional agenda item had been shoehorned in at the last minute. Two representatives of the National Weather Service had driven down from Portland to share some rather alarming new information.

Just months earlier, they explained, scientists had discovered evidence that a fault line in the ocean off the Pacific Northwest coast long thought to be inactive was apparently not so dormant. Scribbling with a grease pencil on an overhead projector transparency sheet, one of the men quickly sketched the ocean floor off Oregon, drawing a line where two tectonic plates—one pushing west, the other pushing east and north—met less than one hundred miles off the Oregon coast. Elsewhere in the world, they explained, convergent plate boundaries such as this one typically produce big earthquakes. This particular boundary was thought, until recently, to be atypically benign. But now geologists had found evidence that this fault line had apparently generated some very large earthquakes in the past, earthquakes of magnitude 8 or greater, and could do so again. Such an earthquake would likely produce a tsunami that would strike the Oregon coast in thirty minutes or less.

Thirty minutes?

A year and a half earlier, Aya hadn't even known enough to worry about tsunamis of any kind. Five minutes earlier, he'd assumed that Cannon Beach would have hours of warning before any tsunami struck the town.

Three days later, at a meeting of the fire district board, Aya shared what he'd learned in Coos Bay. And he made a motion directing the Cannon Beach fire chief to "evacuate hazarded areas of the district upon the occurrence of a perceptible earthquake." Aya wanted to take no chances: if the ground started shaking, the fire district would start evacuating. The rest of the board agreed. His motion was seconded and passed.

This fresh information lent a new urgency to Aya's next task: figuring out where to site the new sirens. The fire district stretched along nearly nine miles of hilly coastline. Aya would have to test the sirens to find out how the sound traveled to determine where to most effectively deploy them. His plan was to mount a siren at the top of a fire truck's cherry picker and drive the truck from place to place, testing the acoustics. But how to activate a siren over and over without causing panic, or at least annoyance, among Aya's fellow citizens? He had strategically named the network of sirens the Community Warning System, or CoWS for short, intentionally omitting the word *tsunami* from the name to suggest that the system could be used for all kinds of emergencies. Maybe, he suggested to a friend, tongue-in-cheek, instead of driving around town sounding a siren, he should test CoWS by blasting a recording of—what else?—mooing cows. Better people should laugh at him than drum him out of town.

Sounds like a good idea, the friend said.

At the public library in Portland, Aya found a BBC recording of a cow mooing in her stall, complete with the clatter of kicking hooves. Back at the fire station in Cannon Beach, he loaded the recording into the system and headed out to test the system's acoustics. Warbling out of a loudspeaker, echoing among the houses and hillsides, the repeated mooing and hoof banging of the peeved bovine was hilarious—more akin to performance art than hazard preparation. In artsy, eccentric Cannon Beach, CoWS was an instant hit.

IT WAS AT A JOINT MEETING of the fire district board and the Cannon Beach City Council that Aya first met Paul Visher. Geologist Curt Peterson, by then a professor at Portland State University, was seeking funding from the city council to expand some paleotsunami run-up research he'd begun doing in the marshes east of town. That meeting, like most such meetings, had been sparsely attended by the public. Only two citizens showed up: a couple in their late sixties. They were newcomers to town. Aya hadn't met them yet, but between the local gossip mill and the *Wall Street Journal,* he knew who they were. Much later, after he and Paul Visher had become close friends, Aya heard what had happened the morning after that meeting. Visher had walked into the Clatsop County sheriff's office in Astoria and, with an authority typically reserved for captains of industry and cabinet members, declared, "You have a problem, and I can help."

By then, Visher was already acquainted with the sheriff. He had recently moved up from southern California, where he had had a long career with Hughes Aircraft, culminating in the position of corporate vice president. He had served briefly as deputy assistant secretary of defense under Robert McNamara during the Kennedy Administration. But about the time he and his wife were planning their move to Cannon Beach, he had run into some legal trouble. Visher had been convicted in federal court of helping to funnel $300,000 in illegal payments from Hughes to officials of an international satellite consortium. Just months earlier, he had been fined $250,000 and ordered to perform 1,500 hours of community service. Given his move to Cannon Beach, that community service was to be performed under the supervision of the sheriff of Clatsop County, Oregon.

Visher was a smart, take-charge guy—he had an engineering degree from MIT and a law degree from Yale—and he quickly immersed himself in the study of geology and oceanography generally and, specifically, the Cascadia Subduction Zone. His interest was not entirely altruistic; he was by then engaged in designing and building an oceanfront house for himself and his wife. It became a monument to tsunami resistance, Visher-style. The house, shaped a bit like the prow of a ship, hung on a frame of not welded but bolted steel beams set in a thick slab of reinforced concrete. Sheathed in impervious pale yellow panels of a kind used in satellite construction, the house looked even to his friend Al Aya like a building better suited to an industrial park, out of place among the cedar-shingled beach houses in the dunes at the north end of town.

Aya had by this time helped the sheriff's office secure a grant to improve tsunami preparedness along the ocean and river shorelines that formed the western and northern boundaries of Clatsop County. The sheriff had asked Aya to lead the planning, but Aya had declined. He didn't mind helping to get Cannon Beach on the right track, but he wanted nothing to do with neighboring Seaside, where, in his estimation, business interests dominated local politics: if it didn't make money, it didn't matter. But Visher stepped up. The topic interested him, he liked running things, and he wasn't bothered by Seaside's or any other town's reluctance to face facts. And he had a lot of community service hours to work down.

The first big test of the county's new tsunami-preparedness measures occurred the morning of March 25, 1993, when people from Seattle to southern Oregon were shaken awake by an earthquake at 5:34 a.m. Among them was Al Aya in Cannon Beach. At first he thought he was dreaming—too much thinking and worrying about earthquakes, perhaps. Then his two-way fire district radio began chattering. He jumped out of bed, got dressed, and was almost to his Jeep when it occurred to him that, if this *was* the Big One, he might not be returning to his house anytime soon, if ever. He went back inside and quickly changed into warmer, more rugged clothes, then hopped into the Jeep and peeled out, leaving a deep rut in his gravel driveway. On his way, he passed the district's fire trucks, on the road and headed to high ground, sirens blaring. Chief Moon had already arrived, unlocking and activating the alarm system, which was warning citizens of "high-speed ocean flooding" and directing them to evacuate to high ground.

Was this it, the Big One? It didn't seem that big—the shaking had lasted less than a minute—but who knew? The fire station was located downtown, just blocks from the beach and steps from the wetlands where Curt Peterson had been coring and finding layer after layer of tsunami sand. Nearly forty minutes passed before the fire station received a teletype message from the state capital indicating the quake's size and location. It was not an offshore quake after all. The epicenter was fifty-four miles south of Portland, making it an intraplate quake, not a subduction zone quake. The magnitude, it said, was estimated at 5.6 on the Richter scale.

So this was not the Big One. And it was not the type of quake to generate a tsunami in any case. Chief Moon sounded the all-clear.

A couple of hours later, Seaside's chief of police and interim city manager, John West, telephoned the Cannon Beach fire station, asking to meet later that morning with Chief Moon, fire district board president Treva Haskell, and Aya, by then the board vice president.

Cannon Beach's decision to evacuate had created problems in Seaside, Chief West explained to the small gathering at the Cannon Beach fire station. The rural fire district's tsunami warning and evacuation order had gone out over the emergency radio system, and people in Seaside and even Astoria had picked up the warning on police scanners in their homes. Calls poured in to the 911 center, overwhelming it, from people wondering why Cannon Beach was evacuating but Seaside wasn't. Even the police department's own communications system got jammed up.

"I've not come here with any intent to criticize," West went on, "but I want to explain why we did not issue an evacuation order in Seaside." The epicenter of the quake had been at the town of Scotts Mills, in the Willamette Valley, and not off the coast, he explained, offering as evidence the teletype he'd received from the state capital that morning—the same one sent to Cannon Beach. The magnitude, he noted, was a mere 5.6: it was not the kind nor size of quake to cause a tsunami.

Aya, Haskell, and Chief Moon listened patiently to West explain Seaside's policy of waiting until a threat has been confirmed before calling for an evacuation. Then Aya asked to see the teletype. He studied it for a moment. Then he looked up at Chief West. "You understand that we did not learn where the epicenter was until"—Aya gestured to a line on the page in his hand—"6:11 a.m.: thirty-seven minutes after the quake." Then he pulled out some pages of his own, displaying them on the table in front of Chief West.

As it happened, Aya was that very week wrapping up a tsunami-mapping project based on new computer modeling by his friend Paul Whitmore, a geophysicist at the Alaska Tsunami Warning Center. Whitmore had undertaken to model the kind of tsunami waves that might strike the Pacific Northwest coast from a hypothetical magnitude 8.8 Cascadia earthquake. Aya had used that modeling and other data to create a series of what he called wave form charts, spelling out projected arrival times, heights, and surge patterns for a Cascadia tsunami at locations all along the Oregon coast. Visher, by then Aya's collaborator, was planning to debut the charts at a meeting of coastal emergency managers in Newport in a couple of weeks.

"I'm sorry I hadn't prepared these earlier and shown them to you," Aya said. "Had I had any idea we might have an earthquake here in the meantime, I would have shared these with you sooner." He then walked West through the chart he had prepared for Seaside, indicating how long it would likely take for the first tsunami surge from a local earthquake to hit Seaside, how high the wave would be, how quickly it would withdraw to a new post-subsidence low, and when a second surge could be expected. Had this quake been the kind to create a tsunami, Aya pointed out, by the time the teletype had arrived at 6:11 a.m. water moving at thirty to forty miles per hour would already have been knee-deep or better at the Seaside Police Station, on its way to an estimated depth of twenty feet within six minutes.

Now it was Chief West's turn to silently study the documents in front of him. His face became visibly ashen. When finally West spoke, it was as if he were speaking more to himself than to the assembled rural fire district officers. "We have a lot more older and disabled people in Seaside," he muttered. "A lot of them are housebound. And nearly everyone has to cross the river and the creek ..."

The meeting didn't last much longer. There wasn't much more to discuss.

Within months, with Visher's help, the Clatsop County sheriff's office had distributed copies of Aya's wave form charts to every homeowner from Astoria to Arch Cape: six versions for six different Clatsop County communities. A graph on the right with two lopsided bell-like curves indicated the estimated water level at the homeowner's location in the first hour after a magnitude 8.8 Cascadia earthquake. First the line dipped, representing the land subsiding and the wave levels at the shoreline coincidentally withdrawing; it rose steeply as the incoming wave crested above the lowered shoreline. The line fell off even more steeply, then bumped back up to indicate an anticipated second surge rolling in. The accompanying text was presented in nine-point type and was full of technical jargon—liquefaction, coseismic subsidence—and ended with an obtuse caveat: "Factors unknown or of different degree than anticipated in computation of these estimates could produce significantly different results."

Written and designed by a statistician in collaboration with a retired aerospace executive, the handout may have been difficult for the average citizen to decipher. But it was something. In fact, on the Oregon coast in 1993, it was all there was.

19

Home

TOM, 1993

BY THE TIME TOM PASSED STALEY'S JUNCTION on U.S. 26 west of Portland, daylight was fading and the pickup's windshield wipers were working overtime, barely keeping up with the rain that fell straight down, as if the skies themselves were defeated by the prospect of another winter. It was early December, a big month for rain in Oregon's Coast Range. A blue tarp covered the boxes filling the pickup's bed; everything else—his grandmother's old walnut table, his headboard and mattress and box springs—was safely out of the weather, crammed into the U-Haul trailer behind the truck. Not that it had been raining when Tom left Spokane, Washington, early that morning, gunning the truck down the freeway on-ramp to squeeze between two westbound semis. Interstate 90 had been dry, rimmed with a fringe of gritty slush, remnant of the last snowstorm, and the flat gray sky rising from the western horizon seemed to promise more of the same. The only bright spot in that cold, leaden sky had been the landing lights of a jet on approach to Spokane International. But Tom was as confident about his packing and tarping skills as he was about the icy conditions he would encounter in the Columbia River gorge, and the rain he'd hit at the base of the Coast Range.

Many times over the previous six years he'd considered moving back to Seaside, just quitting his job and returning to the now-empty house at the mouth of the river. Then he'd get over it, too gutless or too smart or not quite miserable enough to give up a good-paying job. So he'd pick himself up and go wherever it was ASARCO decided to send him next in the American Cordillera and beyond. Five potent words had finally done it.

Extended projects, his boss had said. *In South America.*

That was a combination Tom knew he just couldn't do.

ASARCO's Spokane office had had a job waiting for him when he finished grad school back in 1988. He'd already worked two field seasons for ASARCO in the Rockies: the summer after he'd been in Peru, and again the summer after Bobbie died. Now the company sent him to the Mojave

Desert to look for industrial minerals, non-metallic things like marble and talc. Next he went north, to the foothills of California's Sierra to investigate marble deposits. Then he was off to Ontario in eastern Canada and the northern Appalachian Mountains—more marble—and to a lime plant in Bellefonte, Pennsylvania, where marble's parent rock, limestone, was being processed, and on down the Appalachians, seeking out more industrial minerals in the mountains south as far as Georgia.

His next assignment brought him back to Oregon for a couple of years—Adrian, Oregon, population 147, in the far southeastern corner of the state. He was to investigate the economic potential of a large zeolite deposit that ASARCO was considering mining for use as a whitener for newsprint. Then he was off to the mountains above Culiacan, on Mexico's west coast, and then Vancouver Island and the British Columbia mainland: two months here, two months there.

Home, in those first years with ASARCO, was a company-owned Chevy Blazer stuffed with his clothes and equipment, though he generally slept in motels convenient to wherever he was working that week or that month. His second home was the Kon Tiki Motel, a tolerable and tolerably cheap lodging on North Division Street just a few blocks from ASARCO's Inland Empire headquarters in Spokane. Tom would stay at the Kon Tiki for a few days each December, when he came to town for the annual Northwest Mining Association convention, or whenever he needed to show his face at the home office. He had kept his apartment in Corvallis for a year—it was cheaper than renting a storage unit—and would stop by whenever the season changed and he needed to swap out his wardrobe. He was about two years into the job when his supervisor suggested it was time for Tom to get his own place in Spokane—buy a house, maybe a condo, and stop living on per diem when he wasn't actually in the field and on the clock. So he rented an apartment on Spokane's South Hill. But the apartment never became more than a place to store his stuff, an address to send bills to, a place to crash for a few days between assignments. Tom bought a television and put it on a timer so it would go on and off in the evenings, giving the impression of habitation.

His final assignment for ASARCO had been in Butte, Montana, his base for two years while he explored the country up around Anaconda and Georgetown Lake. By this time, Tom had become ASARCO's golden boy of skarns, or one of them anyway, recognized for his knack for untangling complicated ore bodies and figuring out which way the underground seams and chutes of minerals ran, seams of valuable minerals: copper and gold.

But by 1993, the luster had long since worn off the job with ASARCO, off his entire career as an exploration geologist. It was the color blindness thing again: still his Achilles' heel. Despite all the tricks he'd learned, the work-arounds, Tom knew he would never be able to relax and trust his own senses. As long as he was working with non-metallic rocks—marble, zeolite—or even studying skarns, he could manage, but when mineral prices are up, who wants a geologist who can't find copper and gold when they're right under his feet?

But there was something Tom found even more crippling about prospecting: the loneliness. The job required Tom—a people person—to spend days on end with no company but his own. He began to feel like he was in solitary confinement out in those wide-open spaces. Economic geology had seemed, back in grad school, like the perfect field, and in some ways it still was: intellectually challenging, cubicle-free, often in stunningly beautiful settings. And it was slowly driving him to the brink. He began to live for the sentence or two of greeting he'd exchange with the motel clerk, found himself dragging out his dinner order just to get in a few more words with the waitress, would strike up unnecessary conversations with the gas station attendant. By early 1993 Tom had begun to wonder if his low energy, his difficulty sleeping were perhaps signs of lead poisoning—not an unreasonable hypothesis, given the heavy metals he'd occasionally been exposed to on the job. But blood tests ruled out an organic cause for what he finally had to acknowledge was, simply, depression. It wasn't the minerals that were making him sick. It was the solitude.

Meeting Pam had eased the loneliness somewhat. A mutual friend had introduced them a couple of years earlier, and they'd hit it off right away, easing into an immediate compatibility. They just got along, period: no drama. Tom had never had a girlfriend like that. Pam had a good job at a college in Spokane. She had a daughter from the most recent of her three marriages: a great kid, precocious and no more or less self-absorbed than any kid that age, especially one raised by a doting single mom. Pam had asked Tom to move in with her, but Tom hadn't been ready for that. Nor was Pam ready to pick up and move to Seaside when Tom asked, hypothetically, if she would. Which is where they'd left it when he'd driven away from Spokane that morning. With his long field assignments, Tom never had been around for more than a few days at a time. His move to Seaside wouldn't make much of a difference in the rhythm of their relationship.

It had come almost as a relief when Tom's boss in Spokane had caught up with him by telephone late that fall, explaining that ASARCO wanted Tom to relocate to Denver, site of the company's international division. From there, he said, they planned to start sending Tom to Chile and Peru for extended projects. Apparently someone had noticed something on Tom's resume about research in Peru and "speaks Spanish" (a bit of a stretch, that second part). ASARCO and its subsidiary, Southern Peru Copper Company, had begun doing mineral exploration along the entire length of Chile and in the mountains of Peru, within a couple hundred miles of the Ataspaca prospect. Tom must have seemed an obvious choice.

Tom had been looking for a way out of the job for a couple of years but hadn't found the nerve to quit; now it was easy. Copper exploration? No way was he going to set himself up for that kind of failure again. And if he was lonely now, he couldn't imagine how much lonelier he would be in a place where he knew no one and couldn't even converse with a stranger much beyond *"Muchas gracias"* and *"¿Cuánto cuesta?"* Back in Spokane, he headed to the ASARCO office to negotiate the terms of his departure. It was a friendly separation. ASARCO was reducing its North American workforce anyway and agreed to make Tom a part of that, landing him a severance package.

HOW DO YOU KNOW WHEN YOU'RE HOME? Five years earlier, on his way to Seaside for some R&R after two long, hot months in the Mojave Desert, Tom had found himself pondering that question and—briefly—his own sanity. Heading north on I-5, he had cut over to U.S. Highway 101 near San Francisco, planning to stop for the night in Gold Beach, Oregon. He was about twelve hours into a fourteen-hour drive when he crested Trinidad Head north of Eureka and began the descent toward what a sign identified as Big Lagoon and he got his first good look at the Pacific Ocean. It was February, and the winter waves were riotous beyond the narrow, log-strewn sandbar separating the lagoon from the ocean, but the lagoon itself was still, so still that he could see the V forming in the water behind a great blue heron wading at the water's edge. The cumulous clouds over the ocean were backlit by the setting sun with a kind of gates-of-heaven glory. That's when he heard it. "You're home," a man's voice said, distinctly, a voice similar to his but *not* his. It was not a *thought;* it was a voice, speaking clearly above the engine noise. It was startling and so real that Tom reflexively looked over one shoulder, then the other, confirming that he was indeed alone. So inhabited had the truck felt that he had responded in kind, out loud. "'*You're* home,'" he inquired, "or '*your* home'?" If he was hearing voices, he wanted to get it right. There was no answer. In fact, after two months in the desert, it did *feel* like home: the fog, the gray sea, the spruce and redwood forest interspersed with leafless alders, their ruddy catkins a pink mist against the conifer green.

Tom knew Spokane would never be home. He had briefly considered staying in eastern Washington, looking around for another job in geology, but he never got around to applying for one. He didn't know what he'd do next, but he knew where he was going: to the house he'd grown up in, the house he'd always figured he'd return to, sooner or later.

That house had, predictably, become a source of conflict soon after Bobbie's death. She had left the house to her five children to sort out. David and Jan and their kids were living in a rental house in Seaside; they volunteered to move in and keep the place occupied while they saved up for a home of their own. But David didn't want to remain an owner of the old place, nor did Lynne or Chris. So Tom and Judy had bought out their three siblings.

It seemed, at first, to be a workable arrangement, but it didn't take long for nerves and relationships to begin to fray. Judy felt she had a right to visit and stay in the house she now co-owned, to spend summer weekends there, sleeping in her old bedroom. But that bedroom now belonged to one of David's boys. David balked; it was his home now, not Judy's beach house. And Judy didn't want to invest any money in the improvements Tom envisioned. When Tom finally offered to buy Judy out, it seemed to be the solution Judy was waiting for. Tom became the house's sole owner. David and Jan, for reasons of their own, promptly moved out.

Between field assignments, Tom would sometimes head to Seaside for a few days of maintenance or repairs on the now-empty house, to preserve his investment and maintain the livability of a house he hoped to live in again someday. And admittedly, a piece of it was for Bobbie. Not that the house was any kind of shrine, or that he wanted to make it one. But keeping the place up seemed a simple, honest way to honor her memory. Rather than let it stand empty, he began renting out the house by the week to vacationers, but he never seriously considered granting anyone a long-term lease. It soothed him, while he worked at one lonely, far-flung job site or another, to picture the house in Seaside waiting for him, should he suddenly decide to move back.

As he was now doing: later than he'd hoped, sooner than was prudent, perhaps. He had no idea what he would do once he finished with all the repairs and remodeling projects on his to-do list, nor how he would support himself. But after nearly six years with ASARCO, living mostly on his per diem and keeping his overhead low, Tom had saved up a nice nest egg. Now he would have time to fix the foundation, to return the cottage to a habitable condition, maybe even expand the house and add on a wing. Neighborhood kids had begun fishing off his dock, treating his yard like a public park. It was a good time to re-stake the Horning family's claim in the place.

And it would be an interesting time to be a geologist living in Seaside. Sometimes when Tom stopped in at the ASARCO office in Spokane, he would see a copy of the state Department of Natural Resources' *Washington Geologic Newsletter* lying on a side table in the reception area. Every issue seemed to touch on the Cascadia Subduction Zone and the newly recognized risk of great earthquakes along the coast. Apparently his old grad school friend Curt Peterson was deep into it; Tom had seen Curt's name in the May 1989 issue, talking about the geologically young stratigraphic sequences on the Oregon coast, the "rapid subsidence of marsh deposits as evidence for great subduction zone earthquakes."

Tom did think it odd, however, reading one article after another about the danger of great quakes on the Northwest coast, how infrequently the inevitable tsunami was mentioned. To Tom, whose memory of the 1964 tsunami remained as vivid as if it had occurred the previous week, it seemed that the writers of those articles were so focused on the earthquake that they were neglecting the story's actual climax, the most important chapter of all.

Maybe that was what the voice at Big Lagoon had been about. After all, Big Lagoon sat practically on top of the Mendocino Triple Junction—the offshore meeting point of three tectonic plates and, as such, the southern end of the Cascadia Subduction Zone. Maybe Seaside was, in a certain sense, merely his address, and his *home* was something much bigger.

It was not yet 5 p.m. but already dark when Tom drove past the turn-off to Klootchy Creek County Park, site of the world's tallest Sitka spruce, and merged with northbound U.S. 101. He drove past the floodlit entrance to

Circle Creek RV Park, past the Relief Pitcher Tavern with its steamed-up windows and window boxes fringed with limp, dead geraniums, past the Bell Buoy seafood shop with its neon Dungeness crab. Just past Seaside High School, Tom flipped on his turn signal and veered off the highway. Hard to believe, after more than twenty years away, he was coming back to live, not just to visit. He pulled the truck up next to the darkened house, turned off the engine, sat in the stillness of a late December afternoon for a few seconds. Then he reached for his parka on the passenger seat and opened the truck door. The rain had let up, diminished to a mere heavy mist. Slamming the truck door, Tom headed not inside the house but out where the gravel driveway petered out between the house and cottage, to the dock at edge of the bay, the one he'd begun building for his mom fourteen years earlier, the one his brother David had eventually completed, Jim Jensen's last stand. This time Tom didn't lie down on the dock like he had thirty-five years earlier, when he'd first arrived at the house at the mouth of the river and lay on his belly to look at all the little live things in the water below. It was enough, this night, to stand on the dock and look out across the water, sparkling with the reflected lights from the windows of the houses across the bay above Little Beach, and to drink in the estuary's cold, black, familiar welcome.

20

Freak Accidents

TOM, 1994

TOM WAS RUNNING LATE, AS USUAL, gunning his pickup truck through the yellow light at Warrenton, heading over Youngs Bay Bridge past the glistening mudflats and toward the towering green trusses of the Astoria-Megler Bridge spanning the Columbia. If the light was green at the left turn for the bridge, and if he didn't get stuck behind an RV on the bridge, or if at the end of the bridge that RV turned right rather than left toward the Long Beach Peninsula, he might get to Ocean Park Retreat Center on time. It was Tuesday, October 4: Day Two of a week-long Elderhostel geology course. And if he'd learned one thing on Day One, it was this: Elderhostel students like to start class on time.

Crossing the bridge, Tom mentally reviewed, with an undercurrent of dread, the curriculum he had prepared. The first class had not gone particularly well: he could see it in the students' drooping eyelids, hear it in the silence when he asked if there were any questions. Truth be told, he had bored even himself. It was one thing to teach geology undergrads; they, at least, came equipped with the basics. Most college freshmen—science majors, anyway—knew something about what drove basic Earth processes. But these folks at Ocean Park: they would have grown up in an era when scientists were still thumbing their noses at the theory of continental drift. Where do you even start? Square One—at least, that's what Tom had tried to do: plate tectonics and all the rest. The result was a lecture and slide presentation that was dry as a bone. And what he had prepared for Day Two was, he knew, not much more exciting.

He was passing the old church in Chinook when his attention was drawn to the voice coming through the truck's tinny speakers. It was an urgent message—so said the radio announcer—from the Clatsop County sheriff's office. A magnitude 8.2 earthquake had struck the seafloor east of Shikotan Island, one of the Kuril Islands strung between the Japanese island of Hokkaido and Siberia's Kamchatka Peninsula. The quake had triggered a tsunami that had already hit Shikotan, Hokkaido, and nearby islands and was now headed toward Hawaii, Alaska, and the North American coastline south to California. Certainly southern Washington's Long Beach Peninsula, where he was now headed. And Seaside, which he had just left.

A tsunami warning had been issued for Clatsop County beaches, the announcer went on. Not merely a *watch*, Tom noted with growing excitement, but a *warning*. So the authorities believed that a tsunami striking the coast here wasn't a mere possibility: it was an inevitability. It was projected to strike the Pacific Northwest shoreline at 3:23 p.m. Less than five hours: just enough time, Tom quickly calculated, to drive to Ocean Park, teach the class, take a few questions, and hightail it back to Seaside.

Tom could hardly believe his luck.

EVEN BEFORE THAT OCTOBER DAY, Tom had been busy, busier than he really wanted to be. His plan, upon returning to Seaside, had been to take a break from work, live on his ample savings for a while, and tackle the remodeling projects he'd long dreamed about, had sketched out in his mind on those long days alone in the field. But somehow the months had passed, and the house didn't look much different: some deferred maintenance had been completed, but no new wing, no rehabbed cottage. Hard to say what had happened to the time.

The summer had been salvaged by sedimentology. Curt Peterson had invited Tom to join him for a week of pulling cores at a study site on the lower Columbia River. On a cool and cloudy June evening, Tom had thrown tent and sleeping bag and clam boots into the back of the pickup and driven an hour north and east to Skamokawa Vista Park, base camp for Peterson's field studies that week. By the time Tom arrived, the park campground—a flat mowed field separated from the Columbia River by piles of dredge tailings— was dotted with crew members' dome tents. It was a good-sized group—two groups, in fact, both seeking, separately, evidence of ancient earthquakes in the sediment layers found on the muddy banks of the marshy islands that fit like puzzle pieces along the Oregon and Washington sides of the lower Columbia River. Peterson and a group of students from Portland State University would be pulling and examining cores of the mud and sand and silt layers in the riverbank, looking for evidence of liquefaction: sand dikes that may have boiled up through the mud when an earthquake collapsed the edge of the North American continent like a house of cards, squishing out the water between grains of sediment and squirting water laden with channel sands up through six or a dozen feet of mud to boil out on the riverbank. Scientists had been doing similar work along Alaska's Cook Inlet, looking for the footprints of liquefaction from the 1964 earthquake; Peterson wondered if those same techniques might work for older Cascadia events.

Tom knew he was out of his element that first morning when, after a boat ride from Skamokawa to Hunting Island, he stepped out of the skiff and promptly sank deep into the mud, nearly overtopping his calf-high rubber boots. That's the difference between a sedimentologist and a hard-rock geologist, Tom thought to himself, grinning ear to ear, just happy to be back in the field, in the company of other geologists. Everyone else was wearing chest waders, everyone but Brian Atwater, who was in charge of the second group, consisting of himself and an assistant. Atwater was doing standard penetrometer studies, driving a pipe down into the riverbank with repeated blows of equal force and recording how long it took to penetrate 6 inches of sand—a way of mapping the loose and the compacted layers of sand, part of Atwater's attempt to determine the magnitude of shaking that had occurred during previous great Cascadia earthquakes. Atwater wore his own idiosyncratic get-up: rain pants cut off above the knees—rain shorts— which he wore like chaps over his hip waders, which he had cinched with straps just below his knees to keep the mud from sucking the boots off his feet. Tom had heard of Brian Atwater—as had everyone who was paying attention to geology news out of the Northwest—but hadn't previously met him. Atwater was nearly as tall as Tom, very friendly and very focused and not afraid to get muddy. He had a particular way of crouching in the mud, hunkering down and lowering his center of gravity into a remarkably stable platform of knee and knee, foot and foot. Tom would have liked to schmooze Atwater a little. But Atwater was all business, meticulous about his work and singularly focused, with body language in the field that telegraphed *not now*.

Instead, Tom spent his time chatting up Peterson's students, regaling them with stories from his fifteen years of geologizing on two hemispheres, sharing observations and advice. Tom had never done any kind of sediment coring in salt marsh or riverbank, and he found himself learning nearly as much as the students.

As fall approached, Tom had begun to feel restless. Mornings he usually found himself at Pacific Way Bakery in Gearhart, where he could stretch a cup of coffee into two hours if there was someone to talk to, but by then most of the regulars had already heard Tom's best stories at least once. It was time to start thinking about going back to work. Paul See, a consulting geologist in Gearhart, was nearing retirement and had offered to hand off to Tom an Elderhostel course he had taught every fall for several years running. It paid $50 per class—a three- or four-hour gig, including driving time to Ocean Park. Tom didn't mind the lousy pay at that point; it was something to do and a way to test the waters for teaching. While he was at it, Tom had made inquiries at Clatsop Community College in Astoria, checking to see if they could use his services as a geology instructor starting, say, the following year, fall 1995. So he had been shocked, just a week later, to get a call from a college administrator: classes started Sept. 26, he was told, and "his" Intro to Geology class was fully enrolled.

THE FIRST BONA FIDE TSUNAMI WARNING on the Northwest coast in years, and by sheer luck, Tom happened to be there and not in Butte, Montana, or even Corvallis. Fortunately it was a Tuesday, and his Intro to Geology class for the college met Mondays, Wednesdays, and Fridays. He could stay on the Long Beach Peninsula and watch the tsunami arrive. But the wave's effects would be magnified in the bay at Seaside, and that's where he wanted to be: on the bluff overlooking the bay at the end of Twenty-Sixth Avenue, watching how this tsunami behaved, to match it with his memories of the second-day surge thirty years earlier, to test and even calibrate his childhood recollections. He needed to see the tsunami on home ground. 3:23 p.m.? He should have no problem wrapping up the class and getting back in time.

"Look," he told the room full of gray-haired students after he walked in empty-handed, his slide carousel still on the pickup's passenger seat, "I was going to pick up where I left off yesterday, but we've got a tsunami warning here today. So I'm going to talk to you about tsunamis instead." There were some worried looks, until he explained that this was a *distant* tsunami, nothing to get too excited about as long as they stayed off the beach that afternoon. This time he spoke without notes, describing collisions of tectonic plates, of square miles of water lifted and shoved landward, of waves just a foot or two higher than the sea's surface as they sped toward land, waves that shoaled and crested and crashed upon the shore and, unlike normal waves, just kept coming. He described what he had seen as a boy after a tsunami had traveled from Alaska to Seaside one night thirty years earlier and turned his own neighborhood upside down, busting up bridges and piling cars against houses, his boyish delight at being awake and barefoot outside at midnight, the look of moonlight on sand and seaweed, the tangy odors, seagulls descending to pick off the translucent pink ghost shrimp that lay naked and exposed on the sand-covered lawn.

This time, no one nodded off. He had the class eating out of his hand.

Then he was back in the truck, barreling south through Long Beach and Seaview, skirting Ilwaco and easing through Chinook, back across the Columbia and down U.S. 101 to Seaside and the turn-off that led to his house. He pulled the truck into its usual spot in the driveway, slammed the door and popped into the house—it wasn't quite 3 p.m. yet—to grab his camcorder. Then he strode the two blocks up Twenty-Sixth Avenue to the rise over the bay, the same place where, as a boy, he had carved truck garages out of the sandbank one fine spring day, right above the sand flat where, the next day, he had stood and watched the tsunami from Alaska resurge into the bay, where he and his siblings still gathered every year on the Fourth of July to light fireworks. He stood there alone in his shorts and sweatshirt—it was a warm day—and aimed his tripod-mounted camcorder at the bay mouth, admiring the view and remembering the look of those same tide flats that day in March 1964: the swift spread of water that just kept coming and didn't stop.

Suddenly Tom was no longer alone. Three men appeared on the exposed sand flats below, lugging wooden stakes and a surveyor's theodolite. He watched them with the glee the only kid on the playground feels upon seeing more kids show up to play: whoever they were, they were part of Tom's club. He watched as they plunged the stakes into the sand, clearly intent on measuring the tsunami run-up, and when they were out of stakes, they withdrew to a driftwood log on the bluff a few yards away from Tom, the three of them sitting in a row, binoculars on the bay mouth.

"Waiting for the tsunami?" Tom blurted out after ambling down the bluff, hoping to ignite a conversation. The three pulled away from their binoculars just long enough to smile and nod, then went back on watch. "Well, you picked the best place to see it," Tom volunteered and turned back to the bay mouth himself.

Then they waited. Minutes passed, and nothing out of the ordinary seemed to happen. 3:23 p.m. came and went. And then suddenly—no, nothing. A wave—could that wave be it? Nope, it was nothing. Nothing at all.

Nothing happened.

Tom, who had witnessed a second-day tsunami surge in this very bay from this very spot, who four years later had seen another tsunami's signature in the damp sandbars and hydroplastic bay floor, would know one when he saw one, even a small one. But nothing happened.

It wasn't long before all four of them came to the same conclusion: false alarm. The city had never been evacuated, in fact; the warning had actually been called off shortly after 3 p.m. The tsunami had been, as tsunamis go, not huge—waves six to eleven feet high when it struck northern Japan, diminishing to eighteen inches tops when it hit Hawaii and Alaska's Aleutian Islands. By the time it lapped the shores of Oregon and Washington, it had simply run out of steam. Which was a good thing, given the number of tourists who had not run from but had flocked to the beach at Seaside that afternoon, hoping to witness and snap photos of the big wave.

"Anyone remember the tsunami here in 1964?" Tom asked rhetorically, smiling broadly. He knew the answer; they might have read about it, but they hadn't seen it. Among the four of them, only Tom had seen the immediate aftermath of the first surge and had watched the second-day surge of what was believed, at the time, to be the only tsunami ever to have hit Seaside, Oregon. He had enjoyed sharing his tsunami memories with the Elderhostel students that morning, but it was even more fun with this crowd. By then, Antonio Baptista had introduced himself and what turned out to be two students from Oregon Graduate Institute in Portland, where Baptista was a professor studying coastal processes, focusing mainly on the hydrodynamics of estuaries. His specialty, he explained, was numerical modeling, using computers to simulate and anticipate natural processes.

On Good Friday 1964, Baptista had been a seven-year-old boy living with his family in Angola, at that time a Portuguese colony in southwestern Africa, which explained his soft Portuguese accent. With the country's war of independence then raging, few Angolans may have taken notice of an earthquake in Alaska, much less its spin-off tsunami in Oregon. But Baptista, a civil engineer by training, had developed a particular interest in tsunamis. In just the past two years, he had done field studies in Nicaragua and in Japan following big tsunamis in those two countries. Since 1988, Baptista had been working with staff at the Oregon Department of Geology and Mineral Industries, attempting to use numerical modeling to simulate the action of a Cascadia tsunami on the Oregon coast, anticipating how large it might be and how high it might run up. Which was why Baptista and his students had scrambled to the coast that day: to try to catch an actual, as opposed to a simulated, tsunami in the act, even a small one: to watch it enter the bay and propagate up the waterways in real time.

The next thing Tom knew, Baptista had stuck a tape recorder in front of Tom and had started peppering him with precisely worded questions about the 1964 event. "What do you remember of the tsunami?" Baptista asked Tom. "How fast was the water moving? How high did the water get? Where, exactly?" Tom was happy to respond, filling the tape with descriptions of that second-day surge, his memory of the day still vivid after more than thirty years.

"You know, we are working on doing numerical modeling of tsunamis off the Oregon coast," Baptista explained after he had captured Tom's quick description of the 1964 surge. A tsunami is a complex phenomenon: there is the underwater earthquake that launches it, and the bathymetry of the ocean through which it moves, and the shoreline it collapses upon, and the coastal bays, rivers, and streams it exploits. What Baptista was attempting to do for DOGAMI was to model that process on a computer: to anticipate how fast a Cascadia Subduction Zone tsunami might move and how high it might flood the coast. "But for this, we need to gather information," Baptista said.

"We can do that!" Tom responded. And that's when Tom came up with his Big Idea.

Tom wasn't the only person in Seaside with memories of the '64 tsunami. There were his fourth-grade friends—now, like him, adults in their early forties—and their siblings, and those kids' parents: people who had witnessed the tsunami not only in his own Venice Park neighborhood but out at the Cove or along the Prom, up in Gearhart or down in Cannon Beach. Gather those stories, scour old newspapers for reporters' accounts, put it all together, and you would have something of immense value: a rich database that numerical modelers could use to better understand the behavior of one small but known tsunami, an understanding that might help them refine their computer modeling of a future, certainly much larger, tsunami from a much nearer earthquake. Tom already had an informal collection of photos from 1964: a neighbor's car tipped over and shoved up against a house, a yard stripped of landscaping. Why not solicit the stories that went along with the pictures?

A community meeting, that's what they needed, Tom proposed to the group gathered on the bluff that October afternoon. Invite local people to bring photos and to share their stories on videotape. Tom volunteered to set it up—to reserve a room and publicize it. Two hours ought to be enough time to gather plenty of stories; he could always hold a second meeting, if the response was good.

THE MEETING, HELD AT SEASIDE HIGH SCHOOL on a Tuesday evening five weeks later, wasn't a complete bust. Tom's friend and former neighbor Neal Maine showed up, as did retired police officer Tom McDonald. Baptista had driven over from Portland with a couple of other students for the evening, as had DOGAMI geologist George Priest and some of his colleagues. Tom had set his own camcorder on a tripod to capture the memories of what he had hoped would be a good-sized turnout—certainly more than the handful who showed up. Neal had come at Tom's request; Officer McDonald must have read about the meeting in the previous week's edition of the *Seaside Signal*. "Testimonies and pictures needed for tsunami study," ran the headline over a photo of Clatsop Community College geology instructor Tom Horning, now sporting a brushy mustache, his dark hair neatly trimmed and landing well above his collar.

"I think people conveniently put (a tsunami) out of their minds as a freak accident," the paper quoted Tom, the reporter adding, "He believes the city needs to prepare for 'freak accidents.'"

Talk about indifference, Tom groused to himself, watching the meeting room door hopefully as the clock ticked past 7 p.m. Baptista and Priest didn't seem particularly disappointed by the small turnout; three eyewitnesses were better than none. But Tom couldn't help but take it personally, considering all the people he'd personally urged to attend. At 7:05, he ushered Neal to a spot in front of the camcorder and started the tape. Neal, recently retired from a long teaching career, smiled a little self-consciously, then launched into his version of the events of the evening of March 27, 1964, an evening, he said, that would have been memorable for his family even without a tsunami.

It was near the end of his second year of teaching, he said. He and his wife, Karen, and their two-year-old son lived in Gearhart, on G Street, then called Thirteenth Avenue. The house was just a couple of blocks west of narrow Neacoxie Creek, which emptied into the Necanicum estuary less than two hundred yards to the south. That evening Neal and Karen were actually at Neal's parents' house, a block away, trying to decide whether to head to the hospital with their toddler son, who had split his head open in a fall on the fireplace hearth and might need stitches. That's when Neal heard what he thought was the engine of his neighbor's logging truck starting up. At nearly 11 p.m.? That hadn't seemed right. He had gone outside to investigate and quickly realized that the deep rumble he was hearing wasn't from an engine: it was the sound of dozens of drift logs knocking together, logs that had been carried up little Neacoxie Creek on a fat wave that was flooding over the pavement at Thirteenth Avenue. Clearly the Maine family wouldn't be driving to the hospital—or anywhere else—via Thirteenth, the most direct route. Instead, the whole family piled into a car and headed north through Gearhart to Pacific Way, the tiny town's main street, then east to U.S. 101, and south. They had decided to pass on the emergency room visit and to evacuate to Crown Camp instead, given what was going on. But just before reaching the left turn off U.S. 101 onto Crown Camp Road, Neal realized his tires were no longer on asphalt but on a thick layer of sand. To top off the evening, Neal said—grinning and pausing for dramatic effect—no sooner had they joined the throngs of citizens pouring into Crown Camp, but Karen, then eight months pregnant, began to feel a familiar cramping sensation. It seemed she was going into labor.

What Tom heard: The tsunami surge into the estuary extended north at least as far as G Street. And the surge was still going strong as the wave reached and overtopped the highway, given the depth of sand Neal had observed on the road.

"False alarm," Neal said, concluding his account with another grin: Karen's presumed labor pains subsided in a few hours. Their baby daughter was born a month later. But that wasn't the end of Neal's tsunami-related memories. One year later, the Maines moved to a house in Tom's neighborhood, just on the other side of the tsunami-flattened Finnish Meeting Hall. The house they bought for $3,000 had been damaged by drift logs lodged under the house's pier-and-post foundation. Neal's father, a carpenter, had jacked up the salvaged house and set it on a new foundation. The house creaked and groaned some during that first year until it finally settled onto its new footings. Neal and family had lived in that house for twenty-three years.

What Tom heard: Water high enough to float big drift logs had flooded at least as far south as Twenty-Fifth Avenue. But he already knew that.

Then Tom McDonald stepped in front of the camcorder. His story illuminated the tsunami's behavior in yet another part of town: the beach and river halfway between Tom's house and Broadway. Apparently the Seaside police chief, watching the television news at home, had heard about the earthquake in Alaska, McDonald recalled, but no official warning had been issued for Oregon beaches until shortly after 11 p.m., when the chief had received a cryptic message from the state capital suggesting that a tsunami might strike sometime before midnight. It seemed unlikely to the chief: there had never been an actual tsunami in Seaside, as far as anyone knew. But just to be on the safe side, the chief had posted McDonald and a couple of other officers along the beachfront Prom to keep an eye on things. Officer McDonald was at the end of Twelfth Avenue, at the north end of the Prom, standing outside his squad car alone in the dark, when he heard the familiar low murmur of the surf suddenly increase to an alarming roar. A strong wind rushed out of the west, carrying great wads of sea foam that flew through the black sky over his head. Immediately he had jumped into the squad car and sped east, down Twelfth Avenue. At the Twelfth Avenue Bridge, the last of the six bridges crossing the Necanicum before the river spread out into the broad, shallow bay, he slowed enough to get a look at the river. It should have been full of water at that point—high tide was only about two hours away—but all the policeman could see at the bottom of the river channel was a mudflat, slick with moonlight. Never, not even at the lowest of low tides, had McDonald ever seen the river actually empty out like that. He gunned the engine and sped east over the bridge toward the highway.

What Tom heard: The first surge was apparently preceded by a significant negative phase, a drawdown of water in the estuary and in the creeks and rivers that fed the estuary.

Tom took his own turn in front of the camcorder, reiterating and embellishing the details he had shared with Baptista on the beach five weeks earlier. Then, with no more testimonies to record, Tom began packing up his camcorder. Baptista strode over, smiling.

"Thank you, Tom," Baptista told him. "These testimonies will be very helpful. I'm going to have my student work this up." That was when Tom realized Baptista was waiting for Tom to hand him the videotape. And when, in the same instant, Tom realized he wasn't about to give the videotape to him. Baptista was a big man, squarely built. Tom was a little taller, and now he straightened up to his full height, smiling with a controlled determination.

Over my dead body was what Tom wanted to say. *This presumptuous nabob thinks I'm some kind of coastal nimrod.* So he didn't have a prestigious research position like Baptista: by Tom's calculations, Tom still held the trump card. He was here in 1964. And the videotape was inside his video recorder. "No, I don't think that's a good idea," Tom said evenly, tucking the camcorder into its case with the video cartridge still inside. "We've barely scratched the surface here tonight. There are so many people who remember this event who didn't show up tonight. We've got to gather a lot more testimony from a lot more people." By *we*, it was clear, he meant *I*. "When I'm done," Tom said, knowing he sounded rude, knowing he was being rude, "I'll give it to you."

So he was being territorial, and not very collegial: Tom didn't care. The way he saw it, it was his idea, his project, and he intended to finish it, to take it as far as he could and not just hand it off to some professor and his graduate student from who-knows-where. The testimony of three people was a pretty thin database; Tom knew there remained plenty of memories out there to mine. If people wouldn't come to his meeting, he would simply have to go to them, one by one if that's what it took.

It had been nearly a year since Tom had moved back to Seaside. He still didn't have a full-time job, but now he had a purpose. As a professional geologist and as someone who had witnessed the 1964 tsunami, he knew he was uniquely qualified to gather this information and make sense of it, to give to the computer modelers a comprehensive package in terms they could understand and use: drawdown, run-up, shoaling, and scouring from the primary and secondary and subsequent surges.

He was all that, and something more. He was the barrel-chested ten-year-old boy who, thirty years earlier, had risen before dawn, shouldered a canvas carrier, and delivered the *Oregonian* and *Oregon Journal* to households from the north end of Gearhart south to Eleventh Avenue in Seaside, in storm and in calm, his Schwinn wobbling precariously with each toss of a fat roll of newsprint at someone's front door.

Tom had been the boy who delivered the newspapers. His mom had been everyone's nurse, at one time or another. He knew everyone in town.

21

WANTED

MARSHALL, CALIFORNIA, AND TOKYO, JAPAN, 1994

"Tsunami!" shrieked the people; and then all shrieks and all sounds and all power to hear sounds were annihilated by a nameless shock heavier than any thunder, as the colossal swell smote the shore with a weight that sent a shudder through the hills, and with a foam-burst like a blaze of sheet-lightning.
—Lafcadio Hearn, *Gleanings in Buddha-Fields* (1897)

UNIVERSITY OF MICHIGAN GEOPHYSICIST Kenji Satake was a little surprised by all the fuss. It started at the Geological Society of America's annual conference in Seattle in late October 1994. Satake had simply diverged a bit from the topic of the abstract he had submitted prior to the conference—just enlarged upon it, really—to propose the specific date, time, and magnitude of the last great Cascadia earthquake:

January 26, 1700, at 9 p.m. Magnitude 9.

Next thing he knew, Satake was getting calls from newspaper reporters and even a writer from *Science*. Two weeks later, the *Oregonian* ran a two-page spread on the topic, to the chagrin of Satake's editors at the journal *Nature*, who planned to publish Satake's article—summarizing the process he used to reach that date and that magnitude figure—a year or so later and didn't particularly like being scooped.

Why all the keen interest? Ever since the Monmouth conference in 1987, newspapers in the Northwest had regularly covered emerging studies about the earthquake potential of the Cascadia Subduction Zone. It was hardly news. Other scientists had already done the heavy lifting, using radiocarbon dating and other techniques to narrow the window of time when the last Big One must have occurred; all Satake had really done was to narrow it further. Why was that such a big deal?

It was an Oregon geologist who spelled it out for Satake. "Because," George Priest had said simply, "that makes it real."

THE STORY OF SATAKE'S DISCOVERY BEGAN just a few weeks before the Seattle GSA conference, on a crisp and sunny September morning at a small paleoseismology meeting held at a hillside conference center above Marshall, California, an hour's drive north of San Francisco. Top-heavy Monterey pines and graceful gray eucalyptus trees obscured the view of Tomales Bay, the blue finger of ocean at the base of the hill, but the smell was all coastal California: saltwater and seaweed and the trees' spice. Kenji Satake, on the deck outside the dining hall, glanced at his watch, then zipped up his jacket: still time to get a walk in before the morning's first session convened.

He set off on the Meadow Trail, which quickly broke out of the trees and onto a grassy rise. Ahead he could see a lone figure apparently stopped on the same trail. As Satake approached, it appeared that the man—tall, lank, with a bristle of mustache—was talking to himself, not quietly but at a conversational pitch. That struck Satake as a little odd. Getting closer, Satake recognized him: Alan Nelson, the USGS geologist who had already presented his research on the use of radiocarbon dating to hone in on the date of the last Cascadia earthquake. So not a lunatic, but still ...

"May I ask what you are doing?" Satake asked, smiling an introduction. Nelson smiled back and displayed a hand-held recorder.

"Making a memo."

"Ah!" Satake exclaimed, laughing. Nelson gestured up the trail, and the two men took up walking together.

"I was very interested in your presentation, but I didn't exactly follow it," Satake admitted. "I didn't know radiocarbon dating could be that precise."

"Well, I'm not surprised," Nelson replied. "It's a new technique. We've never used it before." Nelson had been collaborating with Brian Atwater on the Cascadia problem, doing work similar to Atwater's but on the Oregon coast. He was the lead author of a paper he, Wendy Grant, Atwater, Gary Carver, Mark Darienzo, and five other scientists were preparing to submit to *Nature* summarizing radiocarbon evidence that the last great Cascadia earthquake had occurred roughly three hundred years earlier.

From the fjords of southern Vancouver Island to northern California's Mad River Slough, geologists had been plucking samples of long-dead vegetation—peat buried by tsunami sands, and the leaves and stems of plants eventually killed and silted over by high tides along the lowered post-quake coastline—and sending them to radiocarbon labs to determine their age. As Satake knew, such radiocarbon dating has a confidence interval of plus or minus fifty years—meaning the date it provided could be off by as much as a half-century. By averaging these radiocarbon dates, scientists had arrived at a window of time for the last Cascadia quake about one hundred fifty years wide, sometime from the mid-1600s to the early 1800s. David Yamaguchi's tree-ring evidence suggested an even narrower window: 1680 or later. Meanwhile—Nelson explained to Satake as they walked—Minze Stuiver at the University of Washington's Quaternary Isotope Laboratory had pioneered a new high-precision technique specifically to date the bark-bearing roots of quake-buried Sitka spruce stumps. With this new technique, Stuiver was able to narrow the confidence window of radiocarbon dating to plus or minus ten or fifteen years. Specifically, Stuiver suggested the spruce trees had died sometime between A.D. 1690 and 1720.

Satake was impressed—not only by Stuiver's technique but by the narrow timeframe Nelson now proposed.

"So is there anything in the Japanese record about such a quake?" Nelson asked him.

Satake's first impulse was to smile. It was not the first time he had been asked this question. American seismologists were well aware of Japan's extensive historical record documenting every episode of shaking and every flood for more than fifteen hundred years. A good-sized quake off the Oregon and Washington coasts may well have sent a tsunami all the way to Japan. Had such waves been high enough, they surely would have been remarked upon in the journal of a samurai scribe or village headman somewhere on Japan's Pacific coast. Historians at the University of Tokyo's Earthquake Research Institute, where Satake had earned his PhD, were continuing to unearth old documents touching on such events. The anthology they had assembled comprised twenty-one volumes in its most recent edition. That anthology might well hold clues to the date and size of the last great Cascadia quake. But without date parameters to narrow the search, where would you even start looking?

Now, Satake realized, Nelson was handing him those parameters. Maybe the time had come to take a look.

Like Nelson, Satake specialized in the study of earthquakes and the tsunamis they generated. But he was not a dig-in-the-marsh geologist. He was a computer-modeling geophysicist. Satake had grown up mostly on Kyushu, the southernmost of Japan's four major islands. His choice of Hokkaido University for his undergraduate degree was driven mainly by his passion for skiing and hiking and mountaineering, which in turn had led him to an interest in meteorology and glaciology. By his junior year, he had settled on a major in geophysics, prompted in part by the eruption of Mount St. Helens in the Washington Cascades that same spring.

Faculty at Hokkaido University were responsible for monitoring a trio of volcanoes on southern Hokkaido, and Satake began joining their field trips, helping to stretch cable from seismic instruments high on the volcanoes to observation stations a safer half mile or more away and even hiking through snow in wintertime to troubleshoot the monitoring equipment.

Mount Usu had been under particularly close observation since its most recent eruption in 1977. In 1982 it was declared no longer active, but faculty and grad students from the university continued to staff the Usu Volcanic Observatory twenty-four hours a day, keeping close watch. In May 1983, all the university's volcanologists headed to Kyushu for a few days of fieldwork, leaving Satake in charge at the Usu observatory. He was there at his desk on May 26 when, a few seconds before noon, he felt the jolt of an earthquake. His first panicked thought was that the volcano was erupting. But a quick look at the instruments arrayed before him clearly indicated that wasn't the case; a seismic wave had struck Usu and its sister volcanoes in rapid succession, one after the other, suggesting that the source of the ground movement was an earthquake somewhere to the southwest, perhaps in the Sea of Japan. After reassuring the operators of a tourist cable car that ran nearly to the summit of Mount Usu that no eruption was in progress, he turned back to the instrument array, scrambling to keep the monitors supplied with magnetic tape to record the rumble of aftershocks. Then it occurred to him that a quake that large may have caused some damage. He turned on the television to see if there were any news reports.

That was when Satake witnessed his first tsunami.

The quake had occurred in the seafloor off the northwestern coast of Honshu. It took an hour for the tsunami it generated to travel south to the Noto Peninsula, which jutted into the Sea of Japan like the claw of a crab. There, television cameras captured it rolling into the north-facing harbor at Wajima and exploding onto the port with a kind of relentless drive, the churning water pushing boats against a wharf and setting everything it touched adrift. Among the one hundred people killed that day by the tsunami, Satake later learned, were schoolchildren on a visit to the coast from their village in the mountains. They had felt the quake, but their

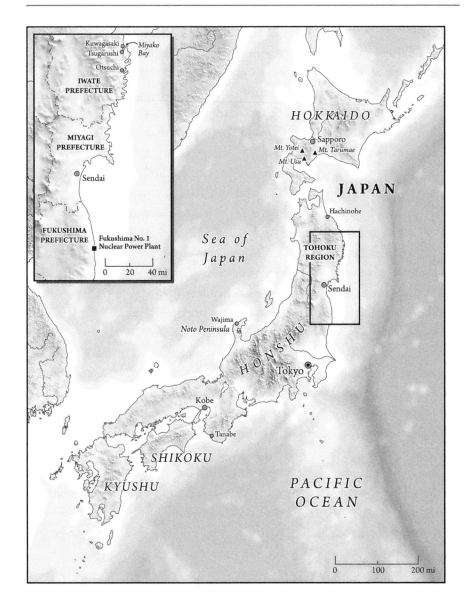

teacher—unaware that a tsunami warning had been issued—decided that the danger had passed, and they had continued on to the beach to enjoy the sunny spring afternoon.

It was, for Satake, a galvanizing event. He turned his attention from a narrow interest in seismology to the relationship between earthquakes and tsunamis. The tsunamigenic characteristics of the Sea of Japan and, in particular, that 1983 Nihonkai-Chubu earthquake became the subject of Satake's PhD research, which he conducted under the supervision of seismologist Kunihiko Shimazaki at the University of Tokyo. It was during a two-year post-doc research position at Caltech in the late 1980s that he first heard about brand-new research on the seismicity of the Cascadia Subduction Zone on North America's west coast.

Now, his interest piqued by the trailside conversation with Alan Nelson, Satake sought out his old advisor, Shimazaki, who was also attending the Marshall conference. Shimazaki was a numbers man, not particularly interested in earthquake history. But he worked closely with two earthquake historians at the University of Tokyo's Earthquake Research Institute and had access to the institute's archives. He didn't really buy Satake's proposition that the last Cascadia earthquake might have been as large as magnitude 9, generating a tsunami that could have traveled all the way to Japan. Magnitude 9 quakes are quite rare; Japan, a country of frequent devastating quakes, had never had one, to anyone's knowledge. Several magnitude 8 quakes seemed, to Shimazaki, a more likely scenario. But he was willing to nose around.

SHIMAZAKI'S OFFICE WAS IN THE LARGER of the Earthquake Research Institute's pair of buildings on the University of Tokyo's Hongo campus, at the north end of central Tokyo. A rank of tall gingko trees, their leaves still green in late September, separated the building from the grounds of three-century-old Nezu Shrine, where a cacophony of bird calls emanated from the tree canopy, filtering into the seismologist's office even through closed windows. It was there, after a couple of days of jetlagged puttering following his return from California, that Shimazaki plucked volumes two and three of the twenty-one-volume *Shinshu Nihon jishin shiryo*—"Newly Collected Materials on Historical Earthquakes in Japan"—from the bookshelf in his office and settled back to read.

Other than the occasional glance to confirm a date or location, Shimazaki had never actually read the anthology. He had left the historical research to two colleagues, Yoshinobu Tsuji and Kazue Ueda. Tsuji, trained in both geophysics and engineering, had joined the ERI seven years earlier after more than a decade with Japan's National Research Center for Disaster Prevention.

Ueda, a petite, soft-spoken woman, had joined the ERI in 1960, shortly after graduating from Tokyo Women's Christian College. But her job as a data-entry clerk for a seismologist took a turn in the early 1970s as a result of a resurgence of radical student activism in Japan. Opposition to the Vietnam War and other issues had led protesters to occupy university buildings, forcing parts of the campus to shut down, sometimes for months or even a full year. Blocked from accessing the ERI's computer, Ueda's boss had turned his attention to historical research—in particular, investigation of original documents pertaining to earthquakes and associated tsunamis. With Ueda's help, he picked up where earlier earthquake historians had left off before World War II, visiting temples and shrines, public libraries and private archives throughout Japan.

There was plenty to look at. Unlike the first people who inhabited the west coast of North America for millennia—people with a rich oral history but no written language—the Japanese had been highly literate for centuries. Earthquake historians in Japan had already documented more than forty-five thousand earthquakes dating as far back as A.D. 416, shortly after the Chinese writing system migrated across the Sea of Japan.

The historical record was especially rich during Japan's Edo Period: the years 1603 to 1868 in the western Gregorian calendar. Literacy was widespread then, not only among the ruling classes but among higher-class commoners, many of whom read and wrote for pleasure, kept detailed financial records, and documented the minutiae of municipal affairs. By the start of the Edo period, Japan had been at peace for a century, and samurai had turned from warriors into bureaucrats. Merchants kept detailed books. Even some peasants read and wrote, for pleasure or instruction or—in their capacity as village officials—to document events of note. Taxes were high and were paid in rice, and records of rice stocks and harvests were meticulously kept. The shoguns who ruled Japan and the daimyo, or lords, under them depended upon samurai and village headmen to collect taxes, explain edicts, keep accounts, and maintain population registers. Both samurai and village headmen tended to use a version of classic Chinese writing in a manner similar to the use of Latin in medieval Europe: it was universally understood by educated people, even if the languages or dialects they spoke were mutually unintelligible.

The classical Chinese brushwork often used in official documents through the nineteenth century could not be understood by most modern Japanese readers, Ueda among them. So she audited university courses in historical document reading. The key, she quickly learned, was to scan the document looking for the word *jishin*, which is expressed in two characters: *ji* for

Earth and *shin* for quake. *Ji* took just a few brushstrokes to form, but *shin* required some fifteen strokes. Its mere complexity made it stand out on a page—that, and the calligrapher tended to make it larger than surrounding characters, just to get all those strokes in. So she would look for that big *shin* in a sea of black brushstrokes. Finding descriptions of high waves or flooding not associated with earthquakes was more challenging. Sometimes the entries were elaborate recitations of events: the time of day, detailed descriptions of the movement of land and waves, accountings of property lost, recitations of local people's reactions, and the writer's own speculation about cause and effect. In other diaries, Ueda might read a simple subject-and-verb entry: *earthquake happened.*

In 1985, Ueda's boss retired, and Ueda went to work for Kunihiko Shimazaki. Given her new supervisor's lack of interest in historical research, Ueda became quite autonomous in her work, and soon she was planning and leading multi-day field trips with ten or more participants to handle and read and photograph old documents found among some family's archives or stored in a temple somewhere. Tsuji, too, did fieldwork, but in his own entrepreneurial style, often dashing out on his own in pursuit of a tip.

Shimazaki knew from Satake that geologists had narrowed the date of the Cascadia quake to within a couple of decades of A.D. 1700. But he decided to read a full century's worth of descriptions. He couldn't read classical Chinese, but he didn't have to; Ueda and others had transcribed the anthology's entries into modern Japanese. He plucked volume two off the shelf, picking up in the mid-1600s and settling in to read everything up to about A.D. 1750.

It took him two full days to read all 699 pages.

Shimazaki was awed. Never before had he read more than snippets of the historical accounts his colleagues collected. Now he found it beguiling: a journey back in time, lyrical descriptions of events he had previously considered only in seismological terms. Immersed in a feudal society intimately tied to a tumultuous land, he was re-experiencing history in the most human of terms: first-person accounts of houses and crops and fishing boats and salt kilns and people lost to shaking and flood and fire.

One unexplained wintertime flood—more than one at about the same time, in fact—caught his attention. But Satake had specified to Shimazaki that the distant tsunami he was looking for had almost certainly occurred in summer. That was the word from Gordon Jacoby, an American geologist who, alongside David Yamaguchi, had been studying tree-ring evidence of earthquake subsidence on the Columbia and other coastal rivers. Jacoby, also at the Marshall meeting, had told Satake that the last Cascadia quake appeared to have occurred just after the spring-summer growth season, suggesting a summer or perhaps fall quake.

"Sorry, no candidate," Shimazaki wrote in an e-mail to Satake. "I found a good one in winter. Sorry it's not in summer." Shimazaki photocopied a few pages of interest from *Shinshu Nihon jishin shiryo* for good measure and sent them to Satake in Michigan. It was a week or two before he heard back from Satake: "Winter is OK!"

Shimazaki was not, he quickly learned, the first to notice a pattern of concurrent and unexplained seaside flooding on the Pacific coast of Japan early in the year 1700. Shortly after receiving Satake's affirmative e-mail, Shimazaki stopped by Kazue Ueda's office for afternoon tea, as was his habit, and told her about the quest Satake had sent him on after the Marshall meeting.

Ueda's eyes lit up. Soon she and her colleague Tsuji were huddled in her office, paging through the bound proceedings of the ERI's third annual historical earthquake conference. They both knew exactly which event Shimazaki was referring to.

The Genroku tsunami.

In 1987, the year he joined the institute, Tsuji had written a scholarly article about a list he had assembled of small tsunamis in Japan's past, tsunamis not necessarily linked to earthquakes, some of mysterious origin. These tsunamis were scattered over two centuries. Only one of them fell in the years 1688 to 1704, the Golden Age of the Edo Period, the era known as Genroku.

Ueda had edited Tsuji's paper; she knew it as well as he did. She and Tsuji now reread the article, focusing in particular on section five, which dealt with what Tsuji had concluded was a tsunami of distant origin that had struck Japan's Pacific coastline on the eighth and ninth days of the twelfth month of the twelfth year of Genroku: January 27 and 28, 1700. Tsuji had begun his narrative by recounting a visit he made, a couple of years prior to 1987, to the city library in the town of Tanabe, about two hundred fifty miles southwest of Tokyo as the crow flies. There he had examined a nearly three-century-old diary written by the mayor of Tanabe, who had reported on "unusual seas" suddenly rising about six feet beginning around dawn on the eighth day of the twelfth month—January 28. Water had flooded a government storehouse and inundated rice and wheat fields, destroying the crops growing there.

Such an event, had it occurred in an isolated area, could have been written off as a storm surge or some other meteorological event. But there was more. According to accounts summarized in the historical earthquake anthology, something very similar had occurred at several sites hundreds of miles away on northern Japan's Sanriku coast—an ancient name for the realm of Honshu's Pacific shoreline stretching from Hachinohe in the far north down to Sendai. There Tsuji had found reports by diarists, all unknown to one another, of inexplicably high waves striking seaside villages many miles apart, apparently on the same winter night. A samurai in the district magistrate's office in the village of Kuwagasaki, on Miyako Bay, noted in his diary that a *tsunami*—the samurai had used that word—had struck late on the eighth night of the twelfth month of Genroku. Villagers managed to escape harm by fleeing to the surrounding hills, but the waves washed away thirteen houses and started a fire—perhaps a lamp or a stove was tipped over, catching a piece of furniture or the wall of a house on fire— that then destroyed another twenty houses.

Wave damage was also reported that night in the town of Otsuchi, several miles down the coast, but it was less severe; two houses were damaged along with fields and rice paddies and two salt-evaporation kilns on the shoreline. The scribe for that town referred to the flooding as a high tide. Neither account made any mention of an earthquake. At both Otsuchi and Kuwagasaki, the flooding occurred at what in the West would be called midnight. (The Otsuchi account mentioned the month and year but not the day—a copyist's omission, Tsuji figured, all other details being aligned.)

The uniformity of the unusual wave heights reported the same day at coastal villages hundreds of miles apart pointed toward a distant tsunami rather than a storm surge, as did the timing, the tsunami having struck the Sanriku coast hours before it arrived at Tanabe, hundreds of miles to the southwest. As did the time of year: in Japan, storm surges typically occur during typhoon season, August to October. In fact, these and other diarists had all reported the weather in Japan on those dates as sunny, or perhaps cloudy, but not the least bit stormy.

Tsuji had noticed another small tsunami reported to have occurred in the town of Tsugaruishi, just south of Kuwagasaki on Miyako Bay, one month to the day before parallel events in Kuwagasaki and Otsuchi. A merchant there had noted that twenty-one houses had been swept away by a high tide that had panicked villagers late at night. That account also touched on the fire that had engulfed houses up in Kuwagasaki. The writer didn't use the word *tsunami*, but he must have thought it; he explicitly noted: "Earthquake did not occur." Tsuji figured a transcription error must have been made in the date; clearly it was the same event.

To Tsuji, all this had added up to evidence of a large tsunami of distant—not Japanese—origin. He referred to the tsunami generated by the magnitude 9.5 earthquake off Chile in 1960 that had struck Hawaii, killing dozens of people, before battering the Japanese coastline with waves more than fifteen feet high; the impact on Japan from that event was strikingly similar to that described in these first-person accounts from three centuries earlier. "From this, I suspect that the origin of the Genroku tsunami might also have been in South America, perhaps off the coast of Chile," Tsuji had concluded. There was no mention of such a quake in existing records from South America, though such records from that era, in that region, Tsuji noted, "are of doubtful accuracy." A North American origin for the Genroku tsunami would not have occurred to Tsuji; in 1987, evidence of the Cascadia Subduction Zone's seismicity had barely begun to emerge.

Three years after compiling the list of small tsunamis, Tsuji had written another paper, this one a plea to fellow earthquake historians to keep an eye out for accounts of distant tsunamis in Japan. "*Shimeitehai—WANTED*," Tsuji had titled the paper, using the word found on posters, in Japan and the United States, seeking dangerous fugitives. With that paper, Tsuji attempted to match known earthquakes around the Pacific Rim and in the Sea of Japan with small-ish tsunamis in Japan. He included a list of known quakes as far back as A.D. 1543 in Korea, South America, the Kamchatka Peninsula, and other sites, linking those quakes with what Tsuji had found in the historical record describing what might have been distant tsunamis striking Japan. Included in the list was brief mention of the Genroku tsunami. Its source, he wrote, was unknown but "probably South America."

Two earthquake history articles published in Japan and dozens of geology papers in American and British scientific journals: as Tsuji and Ueda were puzzling over the origin of what they called the Genroku tsunami, geologists five thousand miles away were using computer modeling, radiocarbon dating, and tree-ring science in an attempt to tease out details of the very same event. And neither group had even been aware of the other's existence.

It was Satake, back at the University of Michigan, who pulled it all together. Reports of waves six to ten feet high in Japan synched with his modeling suggesting a quake of magnitude 9 rather than 8. Satake knew how long it would take a tsunami to cross the ocean: about ten hours from the fault line off the Pacific Northwest to the shores of northern Honshu. There is a seven-hour time difference, as expressed in modern time zones, between the two places. A farfield tsunami striking northern Japan at midnight would have been generated at about 9 p.m. on the Pacific Northwest coast.

Not that anyone was keeping time that way in the plank houses of the villages arrayed on the shoreline an ocean away from Japan. Call it well past sunset on a very dark night—just six days past the new moon, if the moon was even visible in that season of storms.

22

Reunion

TOM, 1995

LONG BEFORE THE MEETING at Seaside High School with Baptista and the others, Tom had begun to assemble an informal collection of photos from the 1964 tsunami: disaster aftermath pictures, mostly by neighbors. There were silky black-and-white images of the bore powering up the river the next day, lapping at collapsed docks, and cheerful Kodachrome snapshots of cars on their sides, of a wooden shed relocated down a sandy road. But with that meeting, and what he saw as Baptista's attempt to horn in on his turf, Tom felt the gauntlet had been thrown down. Tom now intended to gather every written account and photograph, to mine every living memory of the event, and to do what no one else had done: document the experience of one vulnerable Pacific Northwest coastal town with a distant tsunami and, in that process, give something tangible to scientists speculating about where, when, and how fast the next tsunami might flow through town.

He had the time. As New Year 1995 came and went, his only paying job was as a part-time college geology instructor. He started calling up guys he'd grown up with, guys who still lived in Seaside, and soliciting their versions of the event: what they remembered of the tsunami itself and the footprints it had left. Most of the girls—now women—he had known in high school seemed to have fled their hometown, but there were still lots of men around from the SHS Class of '72. He e-mailed those out-of-towners whose e-mail addresses he had. And he had a lot of those; Tom was not one to wait for class reunions to reconnect. He liked staying in touch.

Meanwhile he combed through newspaper archives, reading the accounts of the tsunami in microfilmed pages of the *Seaside Signal, Daily Astorian, Oregonian* and the defunct *Oregon Journal*. The coverage by the local papers, especially, was thorough: any natural disaster is a feast for small-town reporters otherwise bound to recap school board meetings and summarize prep football games. The newspaper photos seemed to plunge Tom back into that down-the-rabbit-hole sensation he remembered from that March day, tooling through the neighborhood on foot and by bike and ogling the devastation. One photo showed two waterlogged cars jammed into a single-car garage. Another showed his neighbors the Jacksons' house split open, one wall hinged at the top like a book falling open. And the photos carried him past the confines of Venice Park to the rest of Seaside and beyond: downtown, to where a workman was photographed scrubbing foam off the First Avenue Bridge, and Cannon Beach, where the bridge crossing what was then called Elk Creek had been ripped from its supports and lay stranded one-quarter mile upstream.

Among the friends who responded to Tom's e-mail was Joe Goetze; they'd been classmates from first grade through senior year. Joe lived in Seattle now and was in touch with lots of former Seasiders. Forget e-mail, Joe told Tom: try using classmates.com. It was brand-new, something called "social networking," as Joe explained it: a way to find and stay in touch with people you knew in school, or anyone you might have lost touch with. (Facebook wouldn't appear for another nine years. Classmates.com relied on paid subscriptions. Tom didn't want to subscribe; he didn't want to provide the personal information required. So Joe made the ask, urging his old classmates to e-mail Tom directly.

Tom also tracked down his friends' parents, and those parents' friends and neighbors, gathering another generation's recollections. He videotaped interviews whenever possible. He saved e-mail messages, took notes during phone conversations. A couple of neighbors wrote up their own accounts of the event and gave them to Tom. On a trip to Sacramento for a land trust conference, Tom snuck in a visit with Charlie and Candy (née Thomsen) Jackson, videotaping them for an hour as they sat on their sofa and shared their individual recollections of that night. Charlie had been seventeen years old in 1964, Candy a few years younger; both of them had lived within a block or two of Tom. They wound up marrying one another and now lived in Folsom, a half-hour from the Sacramento airport. Tom stayed and taped to the last possible minute, throwing the camcorder into the rental car and speeding to the airport and catching his flight with just seconds to spare. He had already tracked down and spoken with Candy's stepfather, Dr. Parcher, the general practitioner for whom Bobbie had once worked. Long retired, he had moved to Florence, halfway down the Oregon coast. "I haven't seen you since I slapped your ass when you were born!" the doctor had chuckled on the phone, although that wasn't quite true; he had also taken out Tom's tonsils, stitched up a couple of gashes, and splinted a broken finger.

Tom found that what some people remembered most vividly were sounds: the diesel-engine rumble of what had turned out, for Neal Maine, to be logs rolling over a road and piling one upon another; a clattering racket, like horses' hooves on a wooden bridge, when the wave caused driftwood to scrape across a gravel road; the jet-engine roar of cobbles and boulders tumbling together in the bed of the river. Others, like Tom, could still conjure the smells from that night: earthy, briny, sulphury odors arising from the disturbed river bottom and mudflat.

There were stories of terror and of coming through in the clutch. The Lowe family, for example: when their house began to fill with water and eight-year-old Steve started yelling, "I don't want to die!" his parents boosted him up onto the roof. There were memories of looting. One neighbor, returning to her house, found a trail of coins outside; someone had apparently come into the flooded house and stolen a coin collection, losing a few coins out of the bag as he beat a hasty retreat. There were stories of selfless bravery and the poignant protectiveness of a fatherless boy. When widowed Takiko Wahl tried to evacuate by car with her two young children, they got caught in the retreating wave. Climbing into the car, nine-year-old Joe was swept away, yelling "Save yourself, mother! I'll be all right!" as he floated, fortuitously, into a shrub of Scotch broom. And there was what some might call selfishness and others simple self-preservation. As Mrs. Wahl and her children flailed in the water, a neighboring family plowed by in a four-wheel-drive pickup truck, hurrying to escape and unwilling to stop and help. "Our children might have died," those neighbors explained matter-of-factly to Tom thirty years later. Mrs. Wahl never spoke to them again. Surprised by an unprecedented tsunami in the middle of the night, Tom wondered, could you really blame them? When pausing to save three more might cause the deaths of twice that many, could anyone say the family in the pickup truck had been wrong?

The project caused Tom to reconsider his own memories, using them to shine a light on the mechanics of a tsunami. Such as the truck garages Tom had carved in the riverbank just hours before the tsunami arrived; they had looked untouched the next day, though clearly they had flooded when the tsunami overtopped the bank. So the tsunami had not been an eroding, scouring wave, Tom concluded. Then how did those ghost shrimp wind up on his lawn, if the riverbed wasn't scoured and the shrimp not scraped out of their holes in the silty sand? Maybe it was the Venturi effect,

Tom speculated: water flowing swiftly by and creating a vacuum effect that sucked the shrimp out.

Ultimately Tom would interview more than seventy-five people. Meanwhile he began giving talks about tsunami risk on the Oregon coast to any group that would have him. The Seaside Downtown Development Association, for one, though they were not particularly receptive to his doomsday projections. He elbowed his way onto the Seaside Rotary Club's list of presenters. One Rotary appearance led to another. A local newspaper reporter heard him speak at one such appearance, and before long, Tom's face, silhouetted by the setting sun and backed by the dunes and surf at Seaside, was on the front page of the *Daily Astorian* over the caption "Tom Horning looks toward the Pacific as another day comes to a close without the Big One." That cutline, and the headline kicker—"Tsounds of Tsunami"—captured the tone of the piece: equal parts informative and ignorant, even dismissive. It focused on a law just passed by the Oregon legislature that would establish a tsunami inundation line along the state's entire coastline, below which most essential public buildings could no longer be built. It consisted mainly of comments by local officials perseverating about how to cope with this new wrench in the works, which threatened to "crimp expansion possibilities." The reporter didn't seem to understand the difference between a distant tsunami, such as the one from Alaska in 1964, and the far larger local tsunami scientists expected the Cascadia Subduction Zone to produce. In fact, nowhere was there mention of the Cascadia Subduction Zone, nor any subduction zone, for that matter.

Not even a *tsubduction* zone.

23

Drawing the Line

HILLSBORO, OREGON, 1995

Earthquake! The ocean roars. Look! Here comes a tsunami.
–Warning inscribed on monument memorializing the 1933
Sanriku tsunami at Misawa, Japan

THE FIRST OFFICIAL OREGON COAST tsunami inundation line—geologists' best guess at how high the water might rise in a typical local tsunami from a typical Cascadia Subduction Zone earthquake—was drawn in 1995. Unless you count Al Aya's wave form charts that the Clatsop County sheriff had distributed to coastal homeowners from Astoria to Arch Cape back in 1993.

But Aya, a statistician, and Visher, an engineer, were amateurs; their estimates of tsunami inundation, while useful, weren't really scientifically sound. And Clatsop County represented only about one-tenth of the state's coastline. The Oregon Department of Geology and Mineral Industries had an entire 362-mile-long shoreline to deal with. DOGAMI's director, Don Hull, had helped organize the Monmouth meeting of geologists in 1987, and it was he who had led transformation of the agency from one focused on extracting natural resources to one with an emphasis on mitigating natural disasters. Now Hull started lobbying the legislature to get serious about setting new seismic standards for building codes in Oregon, not unlike those already in effect in California.

But Oregon had another problem, one most of California did not have. Not only was the earthquake scientists were now expecting off the Pacific Northwest coast likely to be bigger than anything the San Andreas Fault could kick up—an *apocalyptic* quake, as Brian Atwater had described it— but the earthquake rupture zone was underwater. That meant that the Pacific Northwest's next big quake, like those of the past, would almost certainly generate a large tsunami as well. The subduction zone that could cause such a quake stretched from Vancouver Island down to Cape Mendocino, California, just south of Eureka. It was theoretically possible that a tsunami *could* strike the California coast south of Eureka—the San Francisco area, Los Angeles, and the rest of southern California—but it would necessarily be relatively small. An earthquake on the San Andreas Fault could shake hard enough to cause underwater landslides, which in turn could trigger tsunamis that might do some local damage. And all of California was vulnerable to distant tsunamis from big earthquakes far away, the eastern Aleutian Islands, for example. But a distant tsunami would never be a major disaster. With no subduction zone south of Eureka, most of the California coast is immune from a significant *local* tsunami of the kind for which Oregon and Washington and Canada's Vancouver Island had begun bracing.

Hull got to work in 1995 lobbying the Oregon legislature, which—in a rare demonstration of unity and resolve—took his advice and, in one session, passed Senate Bill 379. The bill was designed to keep new hospitals and schools and other key public buildings on the coast from being built in the most vulnerable locations—low-elevation sites likely to be inundated in a local tsunami. How low was too low? The legislature left that to DOGAMI to decide—quickly. By summer 1995 the ball was back in DOGAMI's court. The agency was directed to draw a line to guide new construction along the entire Oregon coast, based on the expected run-up of a local tsunami, and get it done before the end of the year.

The job fell to George Priest, the earnest, soft-spoken staff geologist who led the agency's coastal hazards program. On paper, he was perhaps an unlikely candidate. Priest's doctorate from Oregon State University was in volcanology. He had worked as an exploration geologist and had taught geology briefly at Portland State University. When he joined DOGAMI in 1979, he went straight to work as a geothermal researcher, and he spent the next decade mapping the heat flow of the Earth throughout Oregon's interior. But change, and the ability to flex and retool, had characterized his education and his career path. His undergraduate years at OSU—1967 to 1971—had spanned what was perhaps the most intense period of change in the history of the earth sciences, and he could still recall the wonder he had felt reading Tanya Atwater's seminal 1970 paper on plate tectonics—the sense of everything falling into place. Shortly after the 1987 Monmouth meeting, Priest had become the agency's point man for earthquake hazards in Oregon. But it was the coastal hazards that most interested him. Sure, a big earthquake would likely knock down buildings and bridges and damage roads east to Portland and Seattle and beyond, but it seemed to Priest that the tsunami posed larger risks, certainly to human life. Even before the SB 379 job landed in his lap, Priest had been gearing up. He and Antonio Baptista had traveled together to Nicaragua in 1992 and Japan in 1993 to investigate tsunami run-ups following earthquakes there. And in 1994 he worked with Curt Peterson to launch a study of paleotsunami run-up heights at Siletz Bay, south of Lincoln City, that would soon serve as a guide for drawing a tsunami run-up line for the state's entire coastline.

Siletz Bay was chosen as the site of the pilot project because of its relatively undeveloped shoreline, little changed from what geologists considered prehistoric conditions. The only significant development on the bay was at Salishan, a golf resort and conference center—the very resort where, as a grad student, Peterson had run the slide projector for Vern Kulm at the 1979 Cascadia Margin meeting. It became an international effort, Priest working with scientists from Oregon State University and British Columbia's Pacific Geoscience Centre to design Cascadia earthquake sources for computer-simulated tsunamis. He checked his results against the distribution of paleotsunami sand deposits Peterson had found in the marshes around Siletz Bay and against estimates of the height of Salishan Spit when prehistoric tsunamis struck. His results jibed with the conclusions Satake, Atwater, and others had reached: prehistoric Cascadia earthquakes must have typically been magnitude 9 events.

The timing of the Siletz Bay study was about perfect. The pilot project was wrapping up just as the governor signed SB 379 in September 1995. Now the pressure was on. Assuming a roughly magnitude 9 earthquake, with the fault rupturing the entire length of the subduction zone from Eureka, California, to southern Vancouver Island, with run-ups of about thirty to thirty-five feet, knowing what geologists knew then, or thought they knew, about such events, Priest got to work.

Priest had no training in this area and little research to go by. He set himself up at a desk at Oregon Graduate Institute, tucked away in an office park in Hillsboro, Oregon, in a room with big windows that opened onto a wide lawn edged with a blackberry thicket backed by tall native oaks. From the bright blue days of late summer and through the fall, as the oak leaves crisped and browned and fell, as the days shortened and fog frequently obscured the now-bare oaks, Priest worked seemingly non-stop. Seven days a week, at any rate, and sometimes late into the night, filling the wastebasket with chicken-and-bean burrito wrappers from Taco Bell. His only exercise, it seemed, was to occasionally ascend the open metal staircase outside his office to confer with Baptista, one floor up. He taught himself how to run computer simulations of the seafloor fault rupture process, tutored by experts who were themselves learning new techniques as they went along, and he produced some back-of-the-envelope methods for applying what he projected would happen on the open coast to what would happen when the tsunami ran into bays and up river valleys. It was a heady time for Priest, and not just from the delirium of sleeplessness: here he was, a staff geologist at a state agency, helping to create some of the first-in-the-world computer simulations of earthquakes and tsunamis, for a subduction zone that, just a decade earlier, had been widely believed to be dead or, at worst, harmless.

He was done by Thanksgiving. He had made fifty-six separate maps that, collectively, spanned the entire Oregon coast. Priest knew the maps weren't perfect. They'd been produced on a shoestring and in a tight timeframe (in typical fashion, the state legislature had mandated an outcome without providing any new resources to achieve it). If there were a few places where the lines on the individual maps didn't quite connect, so be it. At least municipalities and school districts and hospitals now had a guide to go by. The line was drawn.

24

WANTED II

Truth to tell, universes revolve within these massive remains. I travel in their systems each time I enter the gravitational field of fascination surrounding every stump.
— Robert Michael Pyle, *Wintergreen*

BRIAN ATWATER YANKED A SECOND TIME on the chainsaw's starter cord and it roared to life, vibrating at the ends of his arms as if the machine itself were eager to get to work. He was crouched neck-deep in a muddy hole nearly four feet deep, poised uncomfortably close to a tangle of stout roots that now lay exposed, awaiting the saw's bite. He and his crew had hand-dug this pit at the base of a tall cedar stump, one of several in the vicinity that had been salvage-logged many years earlier; still evident in the deeply grooved stump's flanks were the springboard notches loggers had used to wedge platforms on which to stand and saw. It stood alone in the salt marsh a short walk from the South Fork Palix River, the next watershed south from the Niawiakum. Nailed just inside a long, V-shaped notch in the stump's otherwise smooth cut face was an aluminum tag the size of a quarter stamped with the number "782."

Atwater braced his back against the pit's mud wall and gingerly lowered the saw's bar tip, watching the whirring, whining chain bite into the base of the fat cedar root, sending wood chips flying. Dead for centuries, the tree's wood still smelled resinous and sweet as a cedar hope chest. The old root was as soft and buttery as a living tree, and the saw sliced through it easily; it took less than a minute for the saw's voice to change, racing up an octave as the severed root dropped with a muffled splash into the mud pooled at the base of the hole where Atwater crouched.

Atwater was no David Yamaguchi, the chainsaw wizard who nine years earlier had found this cedar stump, tagged it, and extracted a wedge of rings from its base. Nor did Atwater have a first-class Stihl like Yamaguchi, but neither was he asking his saw to do the kinds of things Yamaguchi asked of his, sometimes boring into a standing dead tree and, with a modicum of cuts, pulling out blocky cores as long as the saw's bar. Straight cuts from horizontal roots, that's all Atwater needed from his asthmatic old Homelite.

Exactly 296 years and four months: if Kenji Satake was right, that was how long it had been since an earthquake had dealt this tree its fatal blow. The date made sense: an unexplained distant tsunami described in historical documents in Japan aligned with radiocarbon dates procured here in North America. An "orphan tsunami," Atwater had dubbed it. But Atwater wasn't quite ready to accept Satake's figures as fact. The cedars still had to have their say, and with help from him and Yamaguchi, they would.

Atwater set the silenced saw across a couple exposed roots and reached down to pluck the severed slab out of the mud, using his sleeve to wipe away most of the mud. Even raw and unsanded, the old cedar root's varied colors were striking. It was honey and henna in the center, with a wide stripe of ash brown bisecting it; the outer edges appeared as two fat ellipses of ash blonde streaked with orange threads. The root's growth rings swirled and bent like the contour lines on a topographical map, nothing like the ordered concentric circles of a cut tree trunk. Rather, it was a tortured pattern of roots finding their way in soil, seeking water, avoiding other roots. It was a fairly young tree at the time of its death, judging from a quick estimate of the number of growth rings: about two hundred years old, a young adult in redcedar years. The outermost rings seemed to be of roughly equal width: this tree had not wasted away over many years but had died quickly, in a matter of months—slain, one could say, in the prime of life. But what pleased Atwater most of all was seeing this victim's preserved coating of stringy, reddish-brown bark: the tree's skin, perfectly intact.

All in all, a beautiful corpse.

LESS THAN A YEAR AFTER STUMBLING upon the Long Island old-growth cedars, Yamaguchi had moved to Boulder, Colorado, for a job with the University of Colorado's Institute of Arctic and Alpine Research. But he had managed to carve out two weeks the following September to return to the Washington coast and hunt for more cedar snags like those in the Ghost Forest. He and a field assistant drove out from Colorado, stopping in Seattle to pick up the red plastic canoe at his parents' house, before heading to

South Bend, where Atwater had rented a house for the summer. Two weeks wasn't much time, and Yamaguchi wanted to make the most of it. So prior to his trip, he called the local newspapers on the coast to ask for their help.

"Wanted: Dead Cedars" ran the headline in the *Willapa Harbor Herald*. "Dr. David Yamaguchi of the University of Colorado is looking for dead red cedar trees rooted in tidal marshes along the Oregon and Washington coasts." He offered a $20 reward for any tips that panned out.

That notice and one in the Aberdeen *Daily World* yielded several leads. But most of the trees he found the old-fashioned way: paddling the Copalis, the Johns, the South Fork Palix, and other rivers, scanning the marshes and combing the spruce forests for stumps of dead cedars long since logged, searching the treetops for the telltale silver tips of standing snags. In the end, he managed to track down and sample some seventy-five redcedar stumps and snags scattered along fifty-six miles of the Washington coast. By the time he headed back to Colorado, the station wagon's springs were straining under the weight of all those slabs of wood.

Yet the prize he most sought—outermost wood—still eluded him. Fifteen of the seventy-five sampled trees correlated clearly with the Long Island cedars' composite ring pattern. Those trees had lived at least until the 1670s or 1680s, one of them until 1691. But their outermost rings had long since been erased by wind and rain. Exactly how far beyond 1691 they had lived was impossible to tell without those outermost rings.

In spring 1994 Yamaguchi accepted a two-year appointment with the Forestry and Forest Products Research Institute in Sapporo, Japan. He was in his office there one morning in January 1996 when his eye fell on the latest issue of *Nature*. The cover—a full-page reproduction of the nineteenth-century woodblock print "The Great Wave off Kanagawa"— said it all. Never mind that the iconic print actually depicted a storm wave, not a tsunami: Yamaguchi knew right away what the cover story was. He opened the journal and flipped to Satake's article.

January 26, 1700, 9 p.m. Magnitude 9.

The article wasn't a surprise; Yamaguchi had already read an interview with Satake in *Science* about the findings. And he wasn't too disappointed; a tree-ring scientist could never have pinpointed the actual date and time of the last Cascadia event the way earthquake historians in Japan had. Had he found some cedars intact to their final ring, he might at least have named the year of the quake, probably even the season. He still could, if only he had outermost wood.

That's when he got an e-mail from Brian Atwater. Maybe it was Satake's article, maybe it was serendipity, but something had clicked for Atwater: he was pretty confident he now knew where to find outermost wood. Forget looking for intact redcedar trunks, Atwater said. Dig out the roots.

The roots of long-dead Sitka spruce were everywhere on the coast, jutting out of eroding riverbanks, rotten and weathered and, to a tree-ring scientist, useless. But the roots of the cedar snags Yamaguchi had already sampled were still in the ground, standing higher in the marshes. Dig out the roots of some of those same cedars and start cutting out contiguous chunks down the tree, starting where Yamaguchi had extracted his own trunk sample nine years earlier. Proceed down far enough and you'll reach the roots, which Atwater suspected were still sheathed in bark and intact, certainly unweathered. That's where they would find the missing outermost rings: in the roots.

Yamaguchi was nearing the end of his two-year stint in Japan; he didn't plan to return to the States until his visa ran out a few months later. Atwater, meanwhile, was eager to get started cutting root samples. He would need Yamaguchi's files, indicating which redcedars yielded the clearest trunk ring-width data and where on the coast those trees stood. Yamaguchi, facing a deadline to wrap up his financial affairs at the Institute, had already begun shipping his belongings back to the States. Accountants handling his grant funding wouldn't pay to ship personal items to his parents' house in Seattle, but they would pay to ship his research materials back to the Cascades Volcano Observatory, where friends agreed to store his boxes until he got back. That's where Yamaguchi's tree-ring measurement data were when Atwater e-mailed him with his idea: on computer disks stashed in meticulously labeled boxes already sitting in Vancouver, Washington. Staff at the CVO dug out the disks and e-mailed the data to Atwater in Seattle. Yamaguchi's annotated topographical maps, detailing each numbered tree's location, they dropped in the mail.

Atwater's goal was to have a robust collection of contiguous slabs from the most promising of the snags—from weathered trunk wood down to intact root wood—cut and sanded smooth for Yamaguchi when he returned; Atwater's neighbor had already volunteered the use of his floor sander. The ring-width analysis and the data logging would be Yamaguchi's to do.

The results of the project would ultimately be published in *Nature* fifteen months after Satake's own article appeared there. Eight of the twelve snag roots Atwater would unearth and cross-section would, as Yamaguchi would write, provide "reliable estimates of the final year of tree growth." Seven of the eight would clearly show a final season of growth in 1699; hence, they had died sometime after the tree had laid on a thick ring of earlywood in 1699 but before a new ring could start to form in spring 1700. The eighth snag had an additional ten annual growth rings beyond 1699, but that tree stood at an elevation higher than the other seven: not low enough to die a quick death, perhaps, but not high enough above the twice-daily tidal inundation to ultimately thrive.

Satake was right, Atwater and Yamaguchi would agree, according to the cedars slain that winter night in 1700. Tree-ring dates, Yamaguchi would write, "thereby give the 1700 Cascadia earthquake a place in written history even though it predates the region's earliest documents by almost a century."

25

Test-ch'as

This was the time when the earth tipped so far that Downriver Ocean came over the bar and flowed up the river, filling and overflowing the canyon, carrying its waters and its fish and other sea life far inland, past even the Center of the World—farther than it had ever penetrated before.

— Theodora Kroeber, *The Inland Whale*

A COOL, GRAY FALL DAY, THE WIND blowing off the ocean from the northwest, cascading over the headland north of Redwood Creek and eddying restlessly over the shrubby, sandy plain where the creek empties into the Pacific. Standing at the edge of the tule marsh, dressed in chest waders, a gouge corer in his hand, Gary Carver could almost hear the shouts of the stick game spectators circled on the nearby sandbar, almost see the young Yurok men lunging and wrestling, snatching the oak-and-buckskin *tossel* with the hooked stick to pitch it toward the goal line. Their presence was especially palpable in this marsh: shirtless men, straining to run in waist-high water. And if, for a moment, that vision should fade, Carver needed only look up to see Walt "Black Snake" Lara and Butch Marks staring at him from the edge of the visitor center parking lot, intimidating the shit out of him.

Here? Carver gestured, pointing with the tip of his corer. Lara and Marks, unsmiling, arms crossed over broad chests, both shook their heads *no.* They didn't even need to look at one another. Both were big Yurok men. Lara was the bull-of-the-woods; he had worked most of his life falling old-growth redwood trees. Carver was a tall man, but it seemed to him that Lara's forearms were roughly the size of his thighs. A random thought passed through Carver's head: *wouldn't want to face these two in a stick game.* Not that he ever would.

Deborah Carver was there too, but across the highway in the visitor center parking lot, keeping warm in the passenger seat of Yurok elder Blanche

Blankenship's car and chatting amiably with the older woman, her warm expression a stark contrast with Lara's and Marks's impenetrable stares. Deborah had brought a map with her—unnecessarily, but she brought it anyway, as a kind of talisman. She and Gary had long since memorized it: the two old Yurok village sites, one north and one south of Redwood Creek; the locations of the houses that composed the village identified on the map as "Ore'qᵂ," south of the creek; and below the village, the sweat house. They had walked the hill, seen the house sites—now just house pits, overgrown depressions, but recognizable to a discerning eye. Places where Lara's and Marks's ancestors a generation or two back carried out their lives on a bountiful, fogbound coast south of the mouth of the Klamath River. These were the sites of the buildings where the two men's ancestors were surely sleeping when the last Big One struck.

Here? Carver selected another spot, as likely as the last one to catch sand grains suspended in a tsunami sweeping over the bar at the creek mouth and following the path of least resistance to the base of the steep hill. *No,* the large heads shook again.

It's not that Carver didn't appreciate their reluctance to let a geologist trample and poke at what he knew was spirit land, given the tribe's history with white people. But Gary and Deborah had read the "myths," knew which houses were said to have been lapped by waves and which, a little higher on the hillside, weren't, and the story aligned perfectly with estimates Gary had made of the likely tsunami run-up at this latitude, in this topography. He'd checked the typical periodicity of tsunami waves against the amount of time it would take for a man to run from a sweat house at the mouth of Redwood Creek to one at Big Lagoon without getting hit by a second surge, and they jibed perfectly. He really didn't need to core the marsh to know what happened here. But he was a geologist. The opportunity to close this loop to his own satisfaction, squaring the science with the stories, was irresistible to Gary—and to Deborah, the catalyst.

"Mr. Lara, I'm Gary Carver; we've met a couple of times," Carver had begun winningly, after screwing up the courage to knock at the door of the house in the woods with the two snarling dogs outside. The marsh was on Redwood National Park land, but park rangers had told Gary he needed the tribe's permission to core there. The Yurok tribe had sent him to its cultural committee, whose members had sent him to Lara, whose family owned that land. Owned not in a fee-title sense. Owned in an Indian sense.

"No way."

"I would only take a centimeter or two of mud and peat, about a meter deep," Carver said, still standing on the porch, dangling the gouge corer from his fingertips as if it were a gun, open-chambered: *See? Harmless.*

"Nope."

This marsh was, well, Lara hadn't used the word *sacred,* but special: a training ground used by generations of young men preparing themselves for the tribe's traditional stick game, he said, and not to be disturbed.

"I would only take one core ...?" Now Carver was groveling.

"Do you know what happened to my great-grandmother?" Lara quietly inquired.

In fact, thanks to Deborah's research, Carver did know, or thought he did—desecration of her grave, one of many such atrocities—but he didn't say so. That wasn't the point. Knowing the facts wasn't the same as having it happen to you, to your own mother's grandmother.

"Well, if we're there," Lara had finally conceded—*we* being Lara and his cousin, Butch Marks. "And we show you where you can do it. We have to be there."

Here? Carver now gestured again. He had taken several steps to the other side of the marsh and was standing, eyebrows arched, patient, hopeful.

The map Deborah was holding was a page out of *Yurok Geography,* published in 1920 by the University of California Press. She had found it at the Indian Action Council Library in Eureka a few years earlier. Or more accurately, Darlene Magee had found and shared it. Knowledgeable Darlene, who seemed as familiar with the contents of those library shelves as she was with the cupboards in her own kitchen. Wise Darlene, who doled out books to Deborah with a kind of sixth sense, seeming to know what Deborah was ready for before she herself knew it.

Darlene had been the most helpful guide, but many people had helped Deborah follow what had become, unexpectedly, a sort of mid-life quest: a search for expressions of the human experience to match and, in a sense, validate the geological evidence her husband and his colleagues were finding in rock and buried peat and displaced ocean sand. No white people, with their newspapers and diaries and letters home, had been here when a big earthquake last shook the land and pushed the ocean beyond its bounds. Native people surely were, people remembering and passing that knowledge to their children the way they had always shared stories: in wintertime, in living rooms where family gathered, be it a smoky redwood plank house or a three-bedroom, two-bath. Stories that had been told for generations, that were still being told, "myths" that white people hadn't paid any attention to. Until now.

Here?

DEBORAH'S QUEST HAD BEGUN, in a sense, on the drive home from the Monmouth conference in 1987, Gary chattering non-stop about the other scientists he'd met, the discoveries he'd heard about in British Columbia and Washington and Oregon, all of them fitting crazy-perfect with what he had spent the better part of a decade working on near their home in northern California, at the tail end—or the start, depending on how you looked at it—of the Cascadia Subduction Zone. Big earthquakes and tsunamis on the Pacific Northwest coast: there was no longer any doubt, as far as he was concerned.

That infectious enthusiasm he had for his work: it was one of the things Deborah loved about Gary. She, meanwhile, had her hands full running a household with a new baby and a toddler, living on a headland high above the Pacific Ocean in a house made of tree poles sided with shingles they'd cut from old-growth redwood stumps on the property. At least they now had a flush toilet and a propane cook stove—recent improvements. But God help her if the gravity-fed water system developed a hiccup while Gary was out doing fieldwork somewhere.

Deborah and Gary had met in the late 1970s when she was a geology student at Humboldt State. But geology never quite took for her. She had found herself a little overwhelmed by its complexity and by the intellectual firepower of some of her fellow students, not to mention the professors—people who seemed to be able to look at landscape not just as *space* but as *time*. She had decided to take an indefinite break from school and began looking for a place to live, someplace cheap, closer to her weekend job as an interpreter-naturalist at the Redwood National Park visitor center up the road in Orick. One of the geology professors was looking for someone to rent a spare trailer he had parked on his property high on Trinidad Head. So she spent the winter living in Gary Carver's trailer. Within a year she had moved out of the trailer and into Gary's own cabin. Within a couple of years they were married.

All Gary's talk about earthquakes and tsunamis must have been what prompted her to pull out her old copy of *Yurok Myths* one evening after the girls were in bed. A decade earlier, when she was working at the park visitor center, a friend from Humboldt had shown up one day, clutching a copy of the book and babbling about the crazy stories in it, stories about talking animals and characters named Earthquake and Thunder who rambled around the coast wreaking havoc. *Yurok Myths* was actually for sale at the visitor center—she'd probably shelved it herself without really looking at it. So she had bought a copy and read it—skimmed it, actually. It was hard to sit down and just read stories so *out* there, so odd.

But she now found herself paging through *Yurok Myths* nightly, reading snatches before dropping off to sleep. Maybe, she thought, reflecting on what Gary had shared with her on the drive south from Monmouth, the stories weren't really so crazy. Maybe their meaning wasn't opaque at all. Maybe they were actually describing real events, wrapped up in a kind of metaphorical package in an attempt to make sense of otherwise unexplainable events.

"This stuff is just wild," she'd say to Gary, quoting snippets from compelling passages her eyes fell upon: "'Then the ocean began to turn rough (from the anger of the old men). A breaker came over the settlement (of Siwitsu), washed the whole of it away, and drowned everyone ...'

"Or listen to this, this story called 'How the Prairie Became Ocean.' This is a character called Earthquake talking: 'It will be easy for me to do that, to sink this prairie.' Wild, huh? What do you think?"

Had she posed that question to Gary just a few years earlier, he might have said he thought they were nice stories, period. But that was before Monmouth and everything that had led up to Monmouth. What he thought now, what he told her: when a salt marsh—or call it a prairie, if you're an anthropologist seeking the closest English equivalent to a word in a language you barely understand—subsides during a subduction zone earthquake, lowering the shoreline, and that shoreline is inundated by seawater that doesn't drain out even after the tsunami has passed, well, there's maybe no more precise way to describe that process than "prairie became ocean." He thought that the Yurok people had lived on this coast for a long time, and white settlers for, what, a hundred and fifty years? He was a scientist, steeped in the empiricism of the scientific method. But it seemed to him that the Yuroks *were* describing observed phenomena. Maybe in a traditional culture such as the Yuroks', the stories Deborah was quoting were as empirical as it gets.

Deborah's first visit to the Indian library in spring 1989 had been prompted by an item that appeared in the local newspaper one morning. Until then she'd done nothing but flip through *Yurok Myths*, musing about the correspondences between the stories and Gary's own research. Then her eye fell upon a short piece in the Eureka *Times-Standard's* local history column. It was an article that had originally run in the same paper on April 14, 1855.

> THE INDIANS. — *The Indians of this Bay are now holding a*
> *general meeting at the Rancheria on Pattewott River (Mad River).*
> *They are offering sacrifices to the 'good spirit to hold the earth*
> *still': the shocks of the earthquakes lately felt have frightened*
> *them so much that they would have left for the hills if the*
> *"wagas" (whites) had not persuaded them there was no danger.*
> *The council has lasted five days, which time has been spent, in*
> *the day by the old men in handing down tradition, and the night*
> *by the younger ones in dancing, which they keep up all night.*
> *The Bay Indians have a tradition that this Bay was produced*
> *by an earthquake, which swallowed up the land and destroyed*
> *a large and powerful Indian tribe—only a few escaping—which*
> *statement is almost corroborated by the evidences presented to us,*
> *viz: trees buried to the depth of upwards of two hundred feet, and*
> *more palpable proofs in the immense fissures found in the hills*
> *to the southeast of this place, which appear to have been made*
> *within a century.*

There it was again: ancient stories of an earthquake and subsidence, followed by water inundating the land, with the possible addition of catastrophic landslides. And this not from the Yurok people but the Wiyot, an entirely different tribe, speaking a very different language, living twenty or more miles down the coast. Deborah was an avid reader; she'd always liked doing research. Someone ought to look around to see if there were more such stories told by the various tribes native to the northern California coast; why not her?

She started by visiting the Humboldt Room at the HSU library, which housed the local history collection and was run by a friend of hers, but all the materials there seemed to pertain to the white settlers, not to Native people. "You know, there's an Indian library in Eureka," Joan Berman told her. Deborah had never heard of it. It was in Myrtletown, an old section of Eureka east of downtown, and was run by the Indian Action Council. The woman who ran it was named Darlene Magee. "Go talk with her," Joan said. "She can probably help you out."

That first visit had not gone well. Terra was in school, so Deborah left toddler Molly with a friend one morning and drove alone the half hour to Eureka. The library was just as Joan had described it, in a repurposed school district building, sharing a wall with a day care center. It occupied a single small room, with books filling the shelves lining all four walls. Deborah, eager to get started, smiled winningly at the silent librarian and introduced herself.

"You must be Darlene Magee? Joan Berman at Humboldt State suggested I talk with you." Darlene blinked, and waited for Deborah to continue. "I'm Deborah Carver, and I'm curious about the Wiyot people. I understand there aren't very many of them since the massacre that occurred here. I'm interested in learning more about them."

Darlene didn't say anything at first, but her expression shifted slightly. She stared for a few more seconds. "Well," she said at last, "we have a lot of stuff here, and you're welcome to look at it," and went back to her own reading.

Deborah wasn't sure what to make of Darlene. She started scanning the books crammed on the shelves, wondering where on earth to start. Finally she stepped back in front of Darlene's desk and took a new tack.

"I need to know more about the Wiyot people," Deborah began again. "Their culture, their way of life, their myths, those kinds of things."

Darlene paused again before responding. "I think you ought to read some Humboldt history books," she said finally. "Maybe start with Redick McKee."

"Who's that?" Deborah asked. None of this was making any sense. Was McKee an Indian? If she had wanted to read the history of white settlers in Humboldt County, she could have stayed in the Humboldt Room at the university library.

"Well, he's a guy who came here to settle Indian claims," Darlene replied. Deborah didn't have any idea what she was talking about. Perhaps Darlene read that in her expression, so the librarian rose and pulled a book off a nearby shelf, set it on her desk, and pushed it toward Deborah.

"Thanks," Deborah said, forcing a smile. She turned it over; it was a monograph on Humboldt County history—not exactly what she had in mind. But there didn't seem much more to say. She checked out the book, gathered up her things, and left.

Not until that evening, after the girls were in bed, did she get a chance to open the book she'd checked out. Within the first few pages, she began to feel a creeping sense of shame at her own naiveté, her bumbling insensitivity. What she had in her hands, she realized, was not exactly bedtime reading.

Redick McKee, she learned, was a federal Indian commissioner sent to northern California in 1851 to negotiate treaties with the Indians there. He had no knowledge of Native people's culture or their society, nor of the extent of their territory, nor whether the "chief" he was negotiating with had authority to cede any tribal or village lands. But McKee knew that white Californians had no intention of allowing Indians to continue to freely roam their ancestral lands. He and his fellow commissioners managed to negotiate treaties with eighteen tribes that would have dedicated about 7.5 percent of California's total land to Indian reservations. But the proposed treaties ignited a firestorm of protest. Rather than ratify the treaties, Congress voted to round up California's Indians and herd them into miniscule reservations located on the most undesirable lands in the state.

Why had Darlene sent Deborah off to read about Redick McKee? As a nineteenth-century government Indian agent, he wasn't as bad as most. But it was what came next in the monograph that was truly horrifying.

By 1860 not all Indians on the northern California coast had been relocated; several hundred Wiyot people, for example, lived in scattered villages around Humboldt Bay. As more white settlers arrived and claimed land, conflicts with the Native people naturally ensued. Events reached their apogee in the wee hours of February 26, 1860. Sometime after midnight, an organized gang of white settlers from Eureka armed mainly with hatchets, clubs, and knives paddled silently across a narrow channel of Humboldt Bay separating Eureka from Duluwat Island, site of Tuluwat, a Wiyot village, and set upon the sleeping villagers. More than fifty people—or maybe five times that; estimates vary—were murdered on the island that night. And the massacre at Tuluwat village was not an isolated event; it was one in a series of coordinated attacks targeting Wiyot communities all around Humboldt Bay, leaving the tribe decimated. No one was ever prosecuted for the murders, though the identities of the perpetrators were no secret around town. By the end of that year, a white settler had laid claim to what became known as Indian Island, diking it to run dairy cattle.

And that, Deborah learned, was what had happened to the Wiyot people who lived around Humboldt Bay. Their culture. Their myths. Their way of life.

Were any Wiyot left? Was Darlene herself Wiyot? Had Deborah's direct manner, meant to disarm, been presumptuous, rude even? Had she actually insulted Darlene, treating the genocide of a people like nothing more than a curious chapter of local history? The thought of returning to the Indian Library made Deborah cringe.

But she did return the next week, and the week after that. Darlene—a Yurok, it turned out—was never unhelpful; she would rise from her desk periodically, take something off a shelf, and silently hand it to Deborah. Over time, as Deborah—subdued, but persistent—kept showing up, Darlene grew less remote. And between them a kind of mutual respect, if not actual friendship, slowly took root.

At least one in ten people living in the coastal communities from Eureka to Crescent City, it turned out, is Native American—far more than Deborah had realized, or noticed. They are descendants of the Wiyot around Humboldt Bay, or the Yurok, who lived and still live at the mouth of Redwood Creek and all along the Klamath River, or the Tolowa, native to the vicinity of Crescent City and north along the coast as far as Cape Blanco in Oregon, or other tribes. Among these tribes, Deborah learned, it is the Yurok whose language is best preserved and whose oral literature is best known today.

Chief among the early scholars who took a particular interest in the Yurok was Alfred Kroeber. Born in 1876, Kroeber was raised and educated in New York City and was the first person to receive a doctorate degree in the emerging field of anthropology from Columbia University. The Yuroks were the first California Indians he got to know. Between 1900 and 1908 Kroeber collected and preserved, in recordings and transcriptions, a huge trove of Yurok stories, many of which weren't published until after his death in 1960, including those that comprise *Yurok Myths*, published posthumously in 1976.

Yurok Myths is organized in an unusual manner, grouping the stories by the tellers of the tales. One such "informant," as Kroeber called them, was a woman known as Ann of Espeu, Espeu being a Yurok village on the coast between the mouths of Redwood Creek and the Klamath River. "How the Prairie Became Ocean" was the longest of the three stories or story fragments that Ann told.

"This is where it happened. At Sumig," Ann begins, using the Yurok name for what is today known as Patrick's Point. "Thunder lived at Sumig." Thunder and Earthquake, as she tells it, are disturbed that there is no water nearby—water with salmon and hake and perch and seals to feed the people. Then Earthquake has an idea. "I believe I shall do so that the ground sinks," Ann quotes Earthquake. He begins to run around, sinking the ground with his footsteps and spreading water from place to place. But Thunder is underwhelmed: "Pour some more!" he demands. "Every little while there would be an earthquake, then another earthquake, and another earthquake," Ann says. "And then the water would fill those (depressed) places." Ann wraps up by emphasizing the good that came of all this activity: all the food that became available to the people after Earthquake and Thunder brought water to the people in the form of larger bays.

Californians were not unfamiliar with earthquakes; indeed, the great San Francisco earthquake and fire of 1906, which killed some three thousand people, occurred in the midst of Kroeber's research among the Yurok, delaying his summer visit to Humboldt County that year by two months. But in most of California—the part south of Cape Mendocino—earthquakes were known to make the earth split, not sink. The San Andreas is a strike-slip fault, not a thrust fault; it runs mostly on land and is thus not the type to generate a tsunami. In the early 1900s, there was, apparently, no knowledge of any tsunami of note striking the coastline of the United States.*

Several of the most vivid stories in the book were told by a brother of Ann of Espeu, a man who went by the name Tskerkr, or "pitchfork." "Listen to this, Gary," Deborah had said to Gary one evening, looking up from *Yurok Myths.*

"Then it happened that there almost came to be no people (left in the world) on account of (what happened at) this settlement." The story, as told by Tskerkr, was a short one called, simply, "A Flood." As in several similar stories, the events described occurred at night, at the mouth of Redwood

---

*Documents from Washington state suggest that a tsunami from an 1899 Alaskan earthquake was observed as far south as Willapa Bay. Franciscan mission documents and other historical records point to a tsunami following an 1812 earthquake off Santa Barbara, but little was known of it in 1900.

Creek. "For an old man and his brother went into the sweathouse to sleep," Tskerkr said—Yurok custom dictating that men of the tribe sleep not in their family's plank house but in a nearby sweathouse. While they slept, another man entered the sweathouse and "tied their hair together." Again, nothing too unusual: Yuroks sometimes tied their hair together to prepare for a ritual dance. As Tskerkr told it, the men awoke, bumped heads, and got mad, and their anger turned the ocean rough, washing away a village that had stood on the sandbar below.

"Then all the people of Orekw"—the village on the hillside above the creek—"ran off to the top of the hill, wearing their woodpecker-crest headbands: they were afraid." Those would be the deerskin headbands the Yurok called *ma'ak*, emblazoned with the crimson feathers of pileated woodpecker scalps and worn during the Earth renewal ceremony. The ceremony was held on a regular seasonal schedule but could also be done to re-level the world in case of emergencies such as that Tskerkr described.

The story continues: by this time, water was already surrounding Orekw—the opportunity to re-level the world was ebbing. So a man from Orekw ran to the village at Big Lagoon, some ten miles to the south, where there was another sacred ceremonial site. Once there, he reported that people had drowned at Orekw, "but I am afraid the water will cover the whole land," he said. That is when he began to "speak his formula"—in other words, light a fire, offer incense, and speak the words known to make the world right.

"Then," Tskerkr concludes, "the ocean went down."

To anthropologist Alfred Kroeber, his ear tuned to folk-motifs rather than faults, "Flood" and "How the Prairie Became Ocean" were myths, notable mainly for their focus on setting—"the coast and ocean and the shadowy beyond"—and the fondness informants express for that setting. A myth, according to the conventional understanding of the word, is a type of fiction: a traditional story accepted as history but not history per se. Myths deal not in facts but in supernatural beings or heroes whose activities help explain the unexplainable. Anthropologists perceived Ann's and Tskerkr's stories as occurring in "myth time"—the time when the world as the Yuroks understood it was just coming into being, a kind of time-before-time.

But to the Carvers, reading the stories in 1989, it was hard to imagine more succinct descriptions of an indigenous people's response to a great earthquake and tsunami at the coast, at night: an event that had occurred only about two hundred years before Kroeber's and his informants' time, a mere four or five generations removed from living memory.

A footnote and brief description in *Yurok Myths* pointed Deborah to yet another illuminating story: the tale of the inland whale. Kroeber preserved it in another anthology predating *Yurok Myths*, but it was given a second life by his wife, the writer Theodora Kroeber. "You see there what is wrong with the world," Yurok elder Fanny Flounder had told Theodora as the two sat together one sunny day on Fanny's terrace above the mouth of the Klamath River, gazing at a new channel the river had cut through the sand bar. "The earth tips too far and the ocean comes up the river. That is not good. Even whales could come into the river when it is this way."

The story begins on a familiar note. Nenem, a high-born girl, falls in love with a noble but poor boy from the other side of the tracks. Nenem's family banishes him, and he disappears. When her parents realize she's pregnant, they cast her out too. She finds a home with her lover's mother, and in that hovel by the river she raises her son. In the end, after many tests and trials, Nenem's fatherless son grows to become a leader in the community, respected by all. Much of the credit for his success goes to a whale that had been deposited by floodwaters far inland, a whale the boy had stumbled upon one night and who—in body and spirit—would guide him and become his source of strength.

"The story is fiction," Kroeber wrote in her own explication at the end of *The Inland Whale*, her rendering of nine tales drawn from the oral literature of various California Indian tribes into short stories, published in 1959. But Deborah Carver, reading the story just thirty years later, knew things Kroeber didn't know, about tsunamis of a scale that could sweep even a whale from the ocean and deposit it far up the river, farther than the fiercest storm could push. To Deborah's eyes, the fingerprints of a tsunami were all over the page.

IN THE YEARS THAT FOLLOWED, Deborah would visit every library and historical society archive from Ferndale, California, to Coos Bay, Oregon, and beyond, to libraries in Seattle and Berkeley. It was on a visit to the Del Norte County Library in Crescent City, in a file of ephemera about the local Tolowa tribe, that she came across a softbound document titled *Tolowa Tales and Texts*. The author was Pliny Earle Goddard, a missionary when he arrived in northern California in 1897. While there, living among the Hoopa people, Goddard had discovered a passion for linguistics and ethnology, eventually earning a PhD at the University of California and ultimately becoming the leading scholar of American Indian languages of his day.

The booklet, assembled by Alfred Kroeber in the 1950s, was a photocopy of an unpublished manuscript, dated 1902–1911: more than three hundred pages of typed and handwritten text, in English and in Goddard's own phonetic transcription of spoken Tolowa, on lined and unlined paper, graph paper, even university letterhead. *Tolowa Tales and Texts* had no information about the tellers of the tales, nor did it include any scholarly interpretation. It was just the stories themselves.

This was the first collection of Tolowa stories Deborah had come across, and she flipped through the pages with keen interest. The translations were rougher than those in *Yurok Myths*, but the mythic world they conjured was much the same: humans and animals and, sometimes, natural phenomena such as storm winds talking and interacting in ordinary and extraordinary ways in the context of a familiar landscape, a world animated with spirit and possibility and mystery.

Then her eyes lit upon the ninth tale: "The Flood." It was the same topic as Tskerkr's "A Flood," but there the similarity ended, she realized. For one thing, the Tolowa tale's main characters weren't Earthquake and Thunder. They were simply an unnamed teenage girl and her brother. The storyteller placed the events in the fall of the year, apparently at night. The sea was calm. But the people could read the signs: something was coming. And then it happened: "The earth did truly shake from the west and everything on the earth fell down. The water rose in the streams and came up over the banks."

From that point, "The Flood" reads almost like the script of a disaster film. The young man and woman "kept looking back as they ran and saw the water coming from the west." Everyone was running up the mountain— "everyone that lived upon the earth"—but the water was too fast. "All the people and animals were now floating up the mountainside." The water didn't stop rising until, at the brother's instruction, the girl pulled out the woodpecker feather piercing the septum of her nose and stuck it in the ground.

A group of survivors—how many, the storyteller doesn't say—wound up together on the mountain at that point. The young man built a fire and kept it burning all night. "It looked as if there were fog everywhere filling the valleys," the storyteller says. But it wasn't merely fog. When day broke, the people could see that "only the tops of the hills were projecting above the water."

The group stayed on the mountain for ten days, afraid to return home. Finally the young man went down, returning later with reports of "all kinds of creatures both large and small" scattered where the sea had deposited them. He and his sister decided to return together to their home. "But when they came there, there was nothing, even the house was gone. There was nothing but sand. They could not even distinguish the places where they used to live." Everywhere they looked they saw dead animals, dead people, and "soon all the dead ones began to stink." The girl and boy were now utterly alone.

He took off to the south in search of other survivors, walking "entirely around the world." He kept thinking he would find people—maybe other refugees high in the mountains—but in ten days of walking, he found no one alive.

He had hoped to find a wife for himself and a husband for his sister, he told her upon his return. But "[b]ecause there were no people living anywhere, they married." They scraped together a rudimentary home. In time they had a son. They built a better house. They had a second child, a girl. "These two children when they were grown, married. In this way it kept going on till there were many people. They scattered everywhere and in every place there was a man living with his wife."

Deborah was stunned. This was no creation myth with its cast of superheroes endowed with superpowers. Its stark depiction was no less than post-apocalyptic, an end-of-the-world Adam-and-Eve tale triggered by a flood-borne disaster. The story haunted her; it sounded so plausible, so real.

She brought the book home and showed it to Gary, who started babbling with excitement before he'd read to the end of the first paragraph. "This is liquefaction, right here!" he said, jabbing a finger at the page.

Deborah had been a geology student long enough to know what liquefaction was; she took back the booklet and re-read the first couple of paragraphs: "If you sink into the sand and the sea will rise up," the story read. "We nearly sank into the beach.'"

About the only quirky element in the story, in Deborah's reading, was mention, twice, of people caught by the flood turning into snakes. "What do you think?" she asked Gary. "Do snakes maybe symbolize something?"

"Sea cucumbers!" he burst out, and then he told her a story he'd heard from another geologist about the aftermath of the 1964 Alaska earthquake and tsunami, a story from Narrow Cape, at the eastern end of Kodiak Island. The tsunami had flooded a rancher's field there, and when the sea withdrew, it had stranded thousands of sea cucumbers, animals the very shape and—depending upon the species—size of a cucumber. They were common in northern California as well, where they tended to be long—up to sixteen inches—and slender. "A sea cucumber can look a lot like a snake," Gary insisted, "especially if you're looking for a word for something you've never seen before."

But was it really just a myth? A couple of anthropology courses, a few summers on the payroll at Redwood National Park, and a close reading of *Yurok Myths* did not, Deborah was aware, make her an expert. The explications woven through Alfred Kroeber's collection referred to *myths* versus *formulas*—two genres of tales, from a folklorist's perspective, each distinct from the other. She hadn't really grasped the difference. Maybe she was misreading this story, all the stories—reading too much into them. What was it Joseph Campbell had said about the four functions of myth? There was the metaphysical function, awakening us to the wonders of creation; the cosmological function, explaining the shape of our world and infusing it with meaning; the sociological function, passing down moral codes that teach right from wrong; and the pedagogical function, leading us through the stages of life and its rites of passage. Campbell hadn't said anything about a more pragmatic fifth function, call it an *admonitory* function: warning us when to run like hell.

A friend suggested she show the story to Thomas Buckley, a cultural anthropologist visiting Humboldt that semester from the University of Massachusetts. Buckley was not familiar with Tolowa traditions, but he was a student of Yurok spirituality; it was the focus of his academic research. Deborah tracked him down and gave him a copy of "The Flood," asking if he would read it over and give her his impression of what it was: myth, or formula, or something else entirely.

Three days later, Buckley telephoned. "This reads like a newspaper," he told Deborah. "This is not a myth. This is an account." It was, he said, not a creation myth at all, or any other kind of myth. In his opinion, it was a report of an event. Journalism, in essence.

"You need to talk to Loren Bommelyn," Buckley continued. Bommelyn, he explained, was a teacher and linguist and a Tolowa Indian in the Crescent City area. Deborah had heard of Bommelyn; he was well known locally as a traditional basketmaker. And he happened to be the youngest fluent speaker of *Dee-ni' Wee-ya'*—the tribe's own name for themselves and their language.

Buckley had already arranged for Bommelyn to visit his class as a guest speaker the following week. "Come on by," he said. "I'll introduce you." So Deborah did, listening to Bommelyn's lecture and waiting at the end of class for the students to drift away before approaching him.

"I'm Deborah Carver," she began. "I live up in Trinidad; my husband, Gary Carver, is a geologist here. He's been studying evidence of great earthquakes and tsunamis on this coast." Bommelyn smiled politely, listening. "I've been looking for Indian stories from around here that might somehow refer to such events, might describe them or touch on them in some way," Deborah continued. "I found a Tolowa story about a flood that seems quite relevant. I'm wondering if you've ever heard any stories like that yourself, from older relatives, maybe? Do Tolowa still tell any stories like this? Have you yourself ever heard any?"

Bommelyn's gaze shifted slightly, to the classroom wall a little to the right of Deborah's face. He didn't respond right away.

"You know, my parents and my grandmother used to talk about being visited at home by missionaries or pastors, people wanting to talk with them about Jesus, wanting to save their souls, no doubt," he said finally, a sly smile now playing on his face. "They would always talk about the creation story—you know, Genesis—and my mother would say, 'Well, *I* have a story.' And she would talk about this flood, a great flood.

"Mom would talk about how there was an earthquake and a tidal wave, how everybody left and they had to start all over again," he continued. "She used to laugh, telling us the story. The clergymen were polite; they'd say, 'Oh, that's a nice story, Eunice, but a little crazy!' But she'd persist, she'd say, 'This really happened to my people, not very long ago.' Pretty soon they'd leave. They must have thought, *Just another crazy Indian lady.*" Bommelyn chuckled. "You've got to admire their persistence. Both she and my grandmother Alice told this story, this flood story, all the time, especially in the winter, after ten days of rain, when the Smith River would rise into a rage and smash waves on the riverbank below the house. They'd say, 'We had this big earthquake, and we had to start over, because everything was washed away.'"

He'd heard it, then. Loren Bommelyn, a man her age, a Tolowa living in Tolowa homelands in the 1990s, had heard the same story Pliny Goddard had heard from a Tolowa storyteller in 1911. He hadn't read it in a book. He had heard it as a child from his mother and grandmother. Here was a living Tolowa story about a devastating disaster, a tragedy with a kernel of hope and, embedded in it, life-saving instructions for anyone paying attention.

Just how different it is to hear, rather than read, such a story, Deborah learned for herself one balmy night in October 1994. It was Humboldt State University's turn to host the California Indian Conference, an annual affair drawing mostly academics, Native and non-, from throughout the state. Gary and Deborah Carver and their friend Jean Perry, a linguist and student of Yurok, had collaborated on a paper Perry presented in a plenary session summarizing their research linking paleoseismology with the oral literature of northern California coastal Indians. Loren Bommelyn was there too, leading a break-out session on his project to render spoken Tolowa into a keyboard-friendly written alphabet.

The two-day conference wrapped up on a Saturday night with a big dinner in the town of Blue Lake, a fifteen-minute drive up the Mad River from the HSU campus. Conference organizers had enlisted the help of Yurok tribal leaders, who set up an open-pit salmon barbecue behind the old Odd Fellows Hall, threading sides of pink-fleshed salmon on redwood skewers and stabbing them in the dirt around a pit of smoldering alder coals.

Deborah had just sat down to eat when Loren Bommelyn stood up. Bommelyn—dark haired and dark eyed, compact in stature and commanding in presence—smiled slightly as he surveyed the crowd, waiting for his opening. And then he began. "I'd like to tell you a story," he said, "a story I heard many times growing up, a story we call *Test-ch'as: The Tidal Wave*."

Deborah put down her fork.

"Lhaa-'ii-dvn Chit-xu waa 'at-ch'a." *Once then, at the Chetco, it happened.* The Tolowa words rolled off his tongue, staccato and mellifluous, the consonants clicking against one other like pebbles rolling in the surf.

"Daa~-xvt-dvn Chit-dvn yee-yilh-ne ch'ee-sii-ne lhan xvlh-tr'in'-dvn xwii-ta 'aa-ghaa-dvn ..." *It was a fall evening, at the Chetco village, and many teenage boys and girls were acting up ...*

Deborah's eyes filled with tears. She couldn't understand the Tolowa words. But she knew the narrative by heart, could almost track the tale by its rhythm alone. She could see the events in the story unfolding, as if watching a movie, a movie she had previously seen only in black and white. Now, listening to the tale in Tolowa, it was as if she could see it in Technicolor and 3-D.

It was "The Flood," but as Bommelyn had heard it in his own family, with additional embellishments. A grandmother, for one, who urged the young people to "go to *En-may*": Mount Emily, it's called today on maps of southern Oregon. The story Bommelyn had heard from his elders took place not at the mouth of the Smith River, where he had grown up, but at *Chit-dvn*, the mouth of the Chetco River, forty miles to the north. The chilling implication: no Tolowa people on the Smith River survived that tsunami.

Bommelyn had his own thoughts about the story's meaning. It's a powerful parable illustrating the trouble that follows when the laws of the One Who Watches Over Us are broken, he would tell his students and his own three children. The corollary: right living brings blessings. And the story served as an inspiring "template of replenishment," as he put it, for a people who had rebounded even after being beaten to the brink of extinction more than once: by ocean waves, and by waves of white settlers and government agents bent on the tribe's destruction.

It was an analysis Deborah couldn't dispute; there was plenty of metaphor in the story. This very evening, for example: fresh salmon on her plate, the scent of alder smoke in the air, her husband's company in a shared quest more rewarding than she could have imagined at the outset, and now, a front-row seat at the intersection of hard science and human experience, told in Tolowa. She must have been living right.

"Hii-waa-sha~," Bommelyn concluded, speaking the words that always closed Tolowa stories.

That is it.

BY 1996, DEBORAH CARVER FELT she'd completed what she'd set out to do. Gary planned to retire from Humboldt after spring semester 1998, and the whole family was preparing to move to Kodiak, Alaska. Her nearly decade-long search for stories had been so consuming, Deborah sometimes felt it teetered on the verge of obsession. It was time, now, to change gears and to focus on helping her pre-teen and teenage daughters make the move.

Other researchers in the Pacific Northwest had, by this time, discovered other earthquake- and tsunami-themed oral literature from Native people all along the six-hundred-mile-long Cascadia Subduction Zone. Discovered in the Christopher Columbus sense of the word: it had been there all along. Strikingly similar stories told in more than a dozen languages had also been preserved on paper or in recordings or were still being told by people from the Coquille and Tillamook of Oregon, the Duwamish and Snoqualmie of Washington, the Huu-ay-aht of Vancouver Island's outer coast. Some were vague and mythic in tone; others offered concrete descriptions of shaking, or marine flooding, or both. Among them were date-specific stories describing events that had occurred "shortly before the white man's time" or, according to an eighty-four-year-old chief speaking in 1964, "four generations before … my grandfather's father's time."

None of the published literature included stories from the Clatsop Nehalem Indians, whose homelands include the mouth of the Necanicum River at present-day Seaside, Oregon. The Clatsop were among the first Indians in the Pacific Northwest to greet arriving European mariners as early as the seventeenth century. They had entertained members of the Lewis and Clark party when they visited in the winter of 1805–06. The reward for their cosmopolitan attitude was annihilation. The Clatsops, along with the Chinookan people at the mouth of the Columbia River, were among the first Native people in the region to be struck by deadly waves of measles, smallpox, and other diseases against which they had no immunity. Those who survived intermarried with whites or with Indians from other tribes. By 1890, when the first linguist arrived from New York to document the Clatsops' language and stories, it seemed there was no one left who remembered more than a few words and phrases in Clatsop.

Archaeologists, meanwhile, had begun reexamining shell middens in light of geologists' findings. Indians living on Netarts Bay seemed to have survived the last tsunami, or at least continued to occupy the site afterwards; the tall, forested sand spit that kept the wind off Mark Darienzo in 1986 may have helped protect indigenous people from a tsunami in 1700. Brian Atwater found a layer of dense, black charcoal under layers of tidal mud and tsunami sand near the mouth of Oregon's Salmon River—evidence of hearths that had been buried by earthquake subsidence, their fires perhaps extinguished by the tsunami. He also stumbled upon a fragment of twined cedar-root textile buried under a layer of thick mud along the lower Nehalem River: remnant of a basket snatched from a village during a tsunami, perhaps, and deposited on the bay floor by the retreating wave. An archaeologist and geologist working together on the southern Oregon coast's Coquille River found more evidence of earthquake subsidence in ancient fishing weirs placed at stream mouths to catch migrating salmon, smelt, and other fish. They found progressively older weirs—one more than three thousand years old—the farther upstream they searched the river's tributaries. That story aligns with geologists' descriptions of a coast that periodically subsides abruptly, then slowly rises in the hundreds of years between great quakes.

It's impossible to know exactly what happened to Native people living at the mouth of the Necanicum River during previous Cascadia quakes, but it probably wasn't good. Excavations of archaeological sites at the south end of Seaside, conducted in 1987 after state highway engineers proposed straightening U.S. Highway 101, give clear indications that the Clatsop or their predecessors were victims of one or more devastating disasters. Analysis of artifacts—the kinds of houses people built and lived in, the ways they expressed themselves in stone and wood—and other clues suggest that the site was abandoned around A.D. 300, when some kind of drastic change happened to the landscape, and the site was later repopulated by a different group of folks with slightly different aesthetics, different cultural traditions. Later studies by geologist Curt Peterson and colleagues would reveal that Seaside had indeed been struck by a tsunami around A.D. 300—and by four more since then.

Gary Carver finally did find an acceptable site in the marsh at Redwood Creek in which to plunge a gouge corer. It revealed no big surprises: just layers of sand and mud and buried peat, as Gary knew it would. And one small bonus: a tiny black seed, tucked into the topmost layer of buried peat and preserved by the cold anaerobic mud above it. Whether the seed had fallen from the seed head of a clump of sedge or grass growing at that site or blown in from nearby doesn't really matter. What does matter is that this seed would have ripened and fallen to the ground in the summer or fall immediately before the last subsidence event on this shoreline. Before it had had time to sprout, or be eaten, or even rot, the seed had been smothered by a layer of ocean sand and sealed with a layer of thick mud.

Gary sent the seed to the University of Washington's Quaternary Isotope Laboratory and obtained a radiocarbon date: A.D. 1700, plus or minus fifty years.

Not a very remote period at all.

26

Wisdom of the Elders

TOM, 1997

"DID I EVER TELL YOU about my sister, the singing sands story?" Paul See, in the passenger seat of Tom's pickup, window cracked, cigarette wedged between two fingers of his right hand and held, to little effect, near the opening.

"Doesn't ring a bell, Paul." Tom Horning, both hands on the wheel, maneuvering through the curves on U.S. 101 where it rounded Neahkahnie Mountain, heading north. It was a late winter afternoon, the end of the work day, the coast socked in and daylight nearly gone.

"You never heard that? You ever notice how the dry sand at Cannon Beach squeaks when you walk on it, but not at Seaside?"

"Matter of fact, I have ..."

"Great little yarn. My sister—I never told you this? Half sister, actually. I found it in her things after she died, when I moved back to the coast, back in '63, '64—'63, must have been. Just before the tsunami. Wish I'd held on to it."

Tom and Paul had spent the day working at the site of a proposed housing development a few miles south of Cannon Beach. The developer wanted to build several houses on a sloping, forested, ocean-view lot—at least, that was what the developer saw when he looked at the property. Tom and Paul saw that and more: an ancient landslide, part of a huge complex of landslides on the south side of Neahkahnie Mountain, which to them spelled a future landslide. Not something a developer wants to hear. "Destined for destruction" was Tom's cheerful assessment—though the report he would write would take pages to say it, spelling out the site's geologic history, more than anyone would ever read. How soon that next landslide would happen was impossible to predict. The hillside had been stable for at least a hundred fifty years and might hold until the next big earthquake, though it's possible a drenching rain or some other force could overwhelm the inherent strength of the soil. Solid earth: in a geologist's eye, it's as ephemeral as the clouds.

"She was working for, what was it, the WPA? Back in the 1930s." See paused, took a long drag on the cigarette. "That make-work program, the Works Progress Administration, that was it, collecting stories, oral histories. There was a fellow called Duncan—I think that was his last name, I never heard his first name—he hung around Cannon Beach, might have been born there, I don't recall. Clatsop, or part Clatsop, I guess he might have been, or maybe he was Tillamook. Anyway, a real dapper dresser; always wore a derby hat, as she told it. Cecile got him to tell her the story of the singing sands so she could write it down. She was quite a bit older than me; I was just a kid then. I guess it's those tsunami stories you've been collecting, from '64, got me thinking about that story again."

Tom had taken over Paul's geological consulting business after his teaching opportunities began to dry up. Teaching had seemed so promising to Tom, at first. It was a way to promulgate the Gospel of Geology, as Tom saw it: the myriad ways geology influences and even controls our lives. Climate change and the fuels we choose to use. The constraints on where we build our buildings and roads and bridges. Our consumerist lifestyle and what we do with the waste we generate. Preparing for and dealing with natural disasters like earthquakes and tsunamis. But the college wanted its freshman geology course to track with the textbook, and there was never enough time to delve into more interesting topics, nor even to get students outside with pick and shovel, getting dirty and making discoveries for themselves. The college had decided to cut back its geology course offerings anyway, reducing Tom's income from slim to nothing: the pay offered to adjunct instructors had never been enough to live on, and Tom had nearly depleted the savings he'd amassed in his low-overhead years with ASARCO.

The takeover had been a friendly one. "Look, Paul," Tom had spelled out one day to the older man, "I don't think I can keep teaching and losing money like this. I'm going to have to start running competition with you."

"Well, then," See had said simply, "I guess it's about time I retired."

And that was that. Tom, a prospector turned part-time professor, remade himself again, this time into an engineering geologist, reading the land not for the ores hidden within it but for the disaster potential it suggested. Not the most interesting of geology professions, but it was a living, in the place where he wanted to live. Frequently, when he wasn't busy with something else, See still accompanied Tom on consulting jobs. Tom didn't mind having another pair of eyes, and the conversation—always entertaining, with Paul See—helped to shorten the sometimes long drives to and from project sites strung along the coast.

"Yeah, I guess that's what made me think of it, your tsunami stories," See continued. "I must have spent twenty years digging around, trying to find more stories like that, local Indian legends, after this business with the Cascadia Subduction Zone came out. I was planning to write a book, even had a title: *Wisdom of the Elders*. I finally gave it up. Just couldn't make any headway."

The story, as See recounted it to Tom on the drive between Neahkahnie Mountain and Tillamook Head, concerned two large rocks standing side by side just offshore south of Tillamook Head, brother-rocks: Opekwan— Haystack Rock, as it's known today—and next to him, tall and slender, Pewhattie. The two brothers quarreled constantly, sometimes urging storm waters to rise and tear the other apart. Saghalie Tyee, the Creator, finally lost patience with the pair and decided to punish them with a lesson they would never forget. He commanded the ocean to retreat far from shore, exposing the feet of the hostile brothers. Seeing the water retreat and the rocks exposed, the young women of the Killamuk tribe who lived along the shore hurried out to gather *toluks*—mussels—and other shellfish and seaweeds growing on the base of the rocks, stuffing their cedar baskets to the brim. Then Saghalie Tyee commanded the waters to return, and they did, rushing back in with such ferocity that the young women on the rocks had no time to escape. The returning wave of water was so big it spilled over the top of Opekwan, higher than any storm waves, and tore at Pewhattie, cleaving him, scattering pieces of him along the shore and swallowing all the young women. Every family lost daughters. The mourning of the mothers was terrible, wrenching even to Saghalie Tyee himself. To ensure that the people would never forget what can happen as a result of quarreling, Saghalie Tyee embedded the mothers' sorrow in the sands at this place, so that everyone walking on them would hear their cries, and remember.

"Quite a tale, there, Paul …"

"Well, that's the gist of it. I changed a few things, minor things, I can't remember it word for word …"

"Did her story say anything about the earth shaking?"

"No. I noticed that. Nothing about an earthquake …"

"Well, it's not surprising, when you think about it," Tom said. "Earthquakes are scary, but it's the tsunami that kills people. That would be what you'd tend to remember and want to warn people about. I'd love to read the original, Paul."

That, See explained with some regret, was impossible. Apparently the story hadn't met the WPA's standards for historical accuracy—Cecile was supposed to gather oral histories, not tall tales—and as a result had not been archived by the federal government. It had languished in Cecile's own handwritten tablet and had been among her effects, stashed in boxes in a garage on the family homestead, when See had moved back to Gearhart. He'd stumbled across it, read it, then put it back in the box or tucked it away in his office somewhere, he couldn't remember which. He didn't think much about it again until the late 1980s, when he began reading about evidence of giant earthquakes and tsunamis along this very coast. But by then Cecile's notes had disappeared. See spent some time searching for them but ultimately came up empty handed.

"The funniest thing," Paul continued, turning to Tom, crushing out his butt in the truck's ashtray: "You know those Elderhostel classes you took over for me, up in Ocean Park, on the Long Beach Peninsula? I told this story to a class I was teaching there a few years ago, and an elderly lady stood up and started talking about a similar story she had heard as a little girl, growing up down around Brookings, on the Chetco River, near the California border. She was neighbor to some Indian folks, people they used to call the "blue chins" because the older ladies all had blue stripes tattooed on their chins. Anyway, apparently they had a story very similar to the one I had told the class. I'm sure she wasn't making it up.

"I just didn't think much about the story, when I found it in Cecile's things back in mid-'60s. I sure wish I'd followed up on it when I had the chance, years ago. Priceless, that story."

That evening, alone in his mother's old bedroom—now his home office— Tom found himself thinking about Paul's story again. It was a good story, a lesson in comportment and a pragmatic warning: don't bother God with your trivial squabbles. And don't put God in a situation where he might lose his temper or you'll never hear the end of it. This Duncan fellow: if he was middle-aged when he talked with Cecile in the 1930s, he might have heard such a story as a young child around the turn of the century, or earlier. Impossible to know how close See's story was to the story Duncan might himself have heard more than a century ago, in English or Clatsop or Tillamook or maybe Chinook jargon. Assuming Duncan hadn't just made it up. But that seemed unlikely. Knowing what Tom knew now about the region's geologic history, it was impossible to imagine that survivors of such a disaster *wouldn't* perpetuate a cautionary tale about the ocean receding and then roaring murderously back in.

Tom had often tried to picture what it must have been like when the last tsunami struck Seaside. The conversation in the truck with Paul See that afternoon seemed to kindle Tom's imagination, and now he could almost see it. A large plank house, large enough to house several families: that was the standard arrangement on this part of the coast. There had probably been several cedar plank houses not far from Tom's own house, on this spit of land between creek and sea, their presence now memorialized by nothing more than a street name: Indian Way. Explorer William Clark, whose men had set up salt-making operations at the base of Tillamook Head during the Lewis and Clark Expedition's stay at the mouth of the Columbia River the winter of 1805–06, reported encountering Clatsop Indians living in several plank houses clustered in twos and threes around the Necanicum estuary. One hundred and six years earlier, there must have been at least that many plank houses here, houses full of people not much more prepared for calamity than the average citizen of Seaside is today. Nighttime, people safely inside. And rain: usually it is raining here, that time of night, that season of the year.

Fires blazing in both square hearths sunk in the lodge's sand floors, casting erratic shadows on the cedar plank walls. Tom imagined a half-dozen men gathered around one hearth, gambling and singing the familiar high-pitched chants, keeping time with elk-hoof rattles, a clattery sound, like alder leaves in the wind or waves withdrawing from a gravelly beach. A few young women gathered around a second hearth doing handwork by firelight, some coaxing wiry strands of spruce root into baskets and others weaving mats of the kind that, that very evening, draped their shoulders and covered the plank house floors and sleeping platforms, organizing the fragrant dried tule leaves with deft stitches: piercing, pulling, flattening, smoothing. Tom pictured them working silently, maybe humming quietly. The old ones may have been in bed already, reclining under elk skins and sea otter pelts on raised sleeping platforms, some watching the fire and the gamblers through half-lidded eyes and some softly snoring next to heaps of sleeping children, stray arms and legs jutting out from under skin blankets. Dogs would have been lying scattered on the floor, their tails curled around them, nipping reflexively at bony backs popping with fleas.

The rain falling on the high cedar roof would have been barely audible, a soft, familiar thrumming rising and falling with that other water sound, the heartbeat of ocean waves breaking on the nearby beach. Now and then a gust of wind may have driven a few raindrops through a smoke hole or a crack in the gable and stirred the air: close and warm and familiar, laden with smoke and sweat and cedar resin and the scent of oily salmon draped over rafters to dry, the last of the fall run.

The dogs would have heard the first seismic wave before it could be felt, triggering a cacophany of barking, all the dogs rising to their feet at once, dashing from one end of the lodge to the other, yelping in excitement and confusion, as if a stranger had arrived and was walking among them. At the dogs' racket, the young women might have paused in their handwork and looked up, exchanging glances. The gamblers, drugged by the game, probably didn't miss a beat. A baby, startled awake by the din, may have ventured an indignant wail.

Then there would have been a shiver in the floor, a sudden wrenching, and a brief pause before the ground under the floor mats started rolling like the sea, shuddering in waves and pitching the plank house into motion, prompting every man and woman to cower. The walls and rafters would have begun to sway and groan, animating the drying salmon and the firelight filtering through the fish and sparking crazily on the ceiling. Left, right, left, the rafters would have jerked, echoing the undulations of the Earth. The plank house must have seemed like a living thing, as if the posts had taken root and become trees again, swaying in a storm, as if it were the belly of a whale moving in the sea, and the people were the fish inside that whale. Inhuman shrieks would have issued from the house as cedar roof shakes and wall slabs rubbed together, as the ends of the lodge's massive ridgepole twisted in the notches atop the gable end posts holding it aloft. Baskets and cedar boxes piled on sleeping platforms would have begun teetering, rocking, toppling to the floor and scattering their contents—dried berries, cakes of cooked wapato, silvery dried smelt—while salmon rained from the rafters, slabs of gray-pink flesh thumping onto the floor mats. Tom could imagine the dogs, silent now but still on their feet, tails twitching, nervously trotting, noses up and sniffing at the air, ever alert to the appearance of food. No one would have moved toward the oval door hole—walking would have been impossible. They would have stayed put, crouched on the quaking ground, covering their heads with their arms, eyes wide open or squeezed shut, while elders and children curled up tight as fiddleheads on their raised platforms, which moved like canoes at sea in synch with the writhing lodge walls. Babies crying, no doubt, but everyone else silent, cowering and waiting for the upheaval to end. It must have seemed endless, those four, five, six minutes, as if the Earth had decided that, like the ocean, it would now churn unceasingly. Then a sudden popping: a side of oily dried salmon, fallen from the rafter and catching fire.

Eventually the Earth's convulsions would have rolled to a close and the house ceased flailing. Inside nothing would have stirred for a while, save the restless dogs; everyone would have been listening for sounds beyond the drumming of ocean and rain, beyond the babies' wails. The plank house would have been a shambles, food and tools and fur blankets littering the floor mats, a wet wind gusting in through gaps in the walls and roof where planks had been wrenched free. But the house would likely still be standing. Even damaged, it would have been a formidable shelter, strong but flexible, the kind of structure a subduction zone earthquake would spare. Tom imagined the men unfolding from a crouch, straightening up as warriors, eyes darting, searching for an enemy to fight, frustrated to find none, the young women coming to life, reaching for babies, seeking out the injured, gathering up spilled food.

The shaman may have been the first to make a sound. Tom could imagine such a man now deliberately reaching for a stick, beginning to pound it on the wooden hearth: *tump, tump, tump, tump*, a deliberate rhythm, purposeful, commanding. Perhaps he sang, a deep, droning chant, singing the sounds that would set the world right. Other men would have followed suit, finding sticks and joining the shaman in the mournful pounding on wooden hearths, wooden platform posts, drumming on the unbroken bones of the house itself.

Everyone would be silently taking stock, thinking of their own behavior, the behavior of others. Had someone taken a salmon out of season, failed to give proper thanks before a hunt, eaten the wrong food at the wrong time, quarreled indecorously? Then as now, there were rules to be followed, sacrifices to make, debts to pay. The world is full of wonder and of terror, time runs through every life and cycles back, the stories of the dead are reenacted by the living. Never in living memory had Thunderbird rattled not just the heavens but the Earth too. But these people would have heard the stories from a time beyond memory, stories of the Earth jumping. And sometimes in those stories, sea waves came on the heels of that land quaking, great waters rolling up and swallowing all the mussel-gatherers on the beach.

Then a grumbling sound would have risen in the west, like dozens of canoes dragged across beach cobble, like a herd of elk running on hard earth. The sound would have picked up, until it sounded as if the ocean were a roaring fire advancing on the plank house. The people inside would have pricked up their ears. Even the shaman may have paused his drumming, alert. But the Earth would have stayed still even as the sound grew. No one would have moved. Every head would have turned toward the west, every eye opened wide in the flickering firelight, all wondering, worrying, steeling themselves.

In an instant, the west wall of the plank house would have exploded in a slurry of logs and seawater, tree trunks slamming like giant clubs through the wall, splintering it and driving through the house to smash open the other side. Water would have swept up everything in the house, and of the house: women, men, children, babies laced into their cradleboards, dogs, baskets and their contents, roof planks, half-burned firewood, fire itself, its reassuring light. The world would have been nothing but night ocean, churning with sharp edges, suffocating, tumbling, grabbing, drowning. Toward the forest the wave would have rushed, toward it and through it, the branches of the spruces reaching out and snagging passing drift: a cedar bark skirt, a half-woven mat. Onward with its cargo of people and animals and things, finally collapsing at the base of the hills, where the wave would have surged, restless. Then the ocean water that had risen and rolled out of its bounds would have begun to drop and ebb, slowly at first, gaining momentum, finally hurrying almost guiltily back toward the bowl of the sea, seeking shortcuts and carving new channels as it bled back through forest and dune and beach, carrying things that floated: a woman stripped of her skirt, eyes open and unblinking in the rain; splinters and planks of split wood; whole spruce and hemlock trees, their roots scoured of soil; canoes and pieces of canoes; an empty cradleboard; baskets bobbing and twirling in the current; a limp dog. Other things the water would have dumped in place as it withdrew: clean sand; an elk skin collapsed into sodden folds; spear points; piles of long, ropy kelp; fish, gasping; perhaps some dazed and shivering survivors.

High overhead, witnesses would have watched and passed comment: a raven, black as the moonless night, croaking its hollow, mocking call from the upper branches of a tall spruce whose lower branches would by now be festooned with debris, and a drift of gulls, mewing and circling uncertainly, waiting for the sea to withdraw, sea and shore to sort themselves, the land to reassert itself back into the known landscape. Known, yet indelibly transformed.

27

Thirteen

NORTHEAST PACIFIC OCEAN, 1999

No geology without marine geology.

—Philip H. Kuenen

It was a pleasant day, no storms in sight, and no swells large enough to ruffle the R/V *Melville's* rolling progress. A fine July day, warm on deck if you could find a sheltered spot in the lee of the 279-foot research vessel's superstructure in which to soak up the sun. Which some members of the shipboard scientific staff were doing on this transit day, while the *Melville* steamed southeast two hundred miles from Cascadia Channel to Rogue Canyon, forty miles west of Gold Beach, Oregon. There was always something to do aboard a research vessel, but this was a day off from the cruise's primary activity: extracting four-inch-wide core samples of seafloor sediments all along the Cascadia Subduction Zone.

There were no days off for Chris Goldfinger, chief scientist on this research cruise three years in the planning: if he wasn't troubleshooting a ketchup shortage, it seemed, he was consulting with the captain about Canadian immigration clearance. Rarely did his duties actually involve science: Goldfinger would have all winter to scrutinize the cores his hand-picked scientific crew had been pulling from the seafloor, wrapping in Saran Wrap, and stowing in the vessel's refrigerated core locker. In fact, he and Hans Nelson, his collaborator on this study, hadn't done more than glance at the cores so far, one at a time, until they strolled into the *Melville's* main lab fifteen minutes ahead of the presentation Goldfinger planned to make to the vessel's crew that afternoon.

It was a routine part of every such research cruise: a kind of dog-and-pony show, explaining exactly what kind of science the crew was helping to support, what particular questions of oceanography or marine geology the scientists on board were seeking to answer. Goldfinger didn't mind doing it; it was a little like teaching geology to undergraduates. It was always fun to talk science and more fun, in a way, to find ways to explain that science to non-scientists such as the ordinary seamen and able seamen, the oilers and wipers, the engineers and electricians and cooks who kept the R/V *Melville*—largest vessel in the Scripps Institution of Oceanography research fleet—humming along twenty-four-seven. Grad students assisting on the cruise had already done the prep work for Goldfinger's talk, laying out a dozen or so five-foot-long, four-inch-wide sediment cores—now split open to show their stratigraphy—that they'd pulled with the piston corer from the abyssal plain at the base of the continental slope, from locations ranging from the tip of Vancouver Island to the central Oregon coast. All Goldfinger had to do was talk about the cores and what he and Nelson expected they would reveal, and not reveal, about Cascadia's paleoseismicity—the prehistoric record of earthquakes along the Pacific Northwest coast.

He and Nelson had dreamed up this project three years earlier after meeting one another at the Geological Society of America's Cordilleran Section meeting in Portland. Nelson, intrigued by a poster Goldfinger had displayed about his investigations of the seafloor at the Cascadia Subduction Zone, had mentioned a controversial paper Canadian geologist John Adams had published nine years earlier. Well before Brian Atwater and others had found signs of great earthquakes in the coastal marshes of southwest Washington, Adams had proposed that turbidites—undersea landslides—provided proof that ancient earthquakes had rattled the seafloor at the collision of the North American and Juan de Fuca plates. Just the way strong earthquakes can trigger landslides on the continent, Adams had suggested, they probably trigger similar slides on the steep submarine canyons of the continental slope, as close to the source of the rupture site as you could get. Examine the layers of mud and sand in turbidities for sharp contacts—marked changes in the sediment layers, indicating sudden movement such as the shaking of an earthquake—and you can read the history of great earthquakes along the Pacific Northwest, or so Adams proposed.

The concept was appealing, but Nelson didn't buy it, and the more he talked, the more convinced Goldfinger was that the two of them could prove Adams wrong. For one thing, Adams hadn't based his paper on new cores he himself had pulled from the seafloor. He had relied primarily on core logs faxed to him by Vern Kulm, Nelson's former advisor, logs Nelson himself and other students had made of cores pulled with Kulm back in the 1960s.

Adams seemed to have cherry-picked the cores that fit his theory, Nelson said, adding that he himself had other old cores, still stored in the OSU core locker, that he believed contradicted Adams's theory.

The two men's backgrounds were a good fit for the project. Nelson was a sedimentologist thirty years into a career with the USGS. He had deep experience with turbidites in oceans around the world, including the Pacific Ocean off Oregon. Goldfinger had come to Oregon State for graduate school twenty years after Nelson, after earning degrees in both geology and geologic oceanography at Humboldt State; after completing his PhD in structural geology, he had stayed on at Oregon State to do research and teach. Increasingly his work had focused on geophysical studies of the ocean floor at the Cascadia margin.

The essential problem with Adams's theory, as Nelson and Goldfinger saw it: it would take all the fingers on two hands to count off all the different things besides earthquakes that could create turbidity currents that might travel to the Pacific Ocean's deep abyssal plain. Storm waves were the most likely culprits, especially close to shore. Tsunamis triggered by earthquakes from near or far might do it. And it didn't necessarily have to be anything catastrophic; sediment accumulating over centuries, stacking up steeply, might eventually just collapse when its angle of repose is exceeded. Volcanic explosions, such as that of Mount St. Helens in 1980, might rock the Earth's crust enough to cause slope failure. Sea level change, meteor impacts, melting of methane-mud ice: the list went on. Would it even be possible to sort out, and rule out, all those potential sources of turbidites?

Well, Goldfinger figured, start by not using old cores pulled from random sites on the continental slope. Rather, pull fresh cores at carefully selected sites, the kinds of sites least likely to have been affected by, for example, disturbance from storm waves. Places like Barkley and Juan de Fuca canyons, nearly a hundred miles off Vancouver Island, and Astoria Canyon and the deep-sea Cascadia Channel as much as one hundred sixty miles off Washington and Oregon, as well as Rogue Canyon and similar canyon systems off northern California. Next, closely examine the sediments in the cores to get their stories. And, most telling of all, look for synchronicity in those cores. Other than a meteor impact—and those were few and far between—only a giant earthquake would trigger undersea landslides all along the Pacific Northwest coast, from Vancouver Island to northern California, all at the same time.

Would that Adams were right, Goldfinger had mused. If assembling a calendar of past great earthquakes along the Cascadia Subduction Zone were as simple as counting turbidites in twenty-five feet of undersea mud, you wouldn't need Chris Goldfinger, PhD, to run a research cruise like this, except maybe to write the grants and oversee distribution of condiments. A five-year-old could do it.

Which was what Goldfinger was still thinking when he met Nelson in the main deck lab and began making mental notes for his talk. Crew members had already started to drift into the lab for his presentation. Now Goldfinger began to browse among the long lab tables, looking at the split cores side by side for the first time. The students had poked markers into some of the cores to point out key contacts—places, for example, where a layer of larger sand grains gave way to a layer of fine olive-gray silt or to what Goldfinger and Nelson both recognized at a glance as Mazama ash. Layer 0, as it was known among geologists: the tephra that shot into the air during the great eruption of Mount Mazama in southern Oregon in 5677 B.C., give or take one hundred fifty years. That eruption blew one vertical mile off the mountaintop into the air and left a deep caldera that eventually filled with rain and snowmelt to become known, in the nineteenth century, as Crater Lake. Ash and pumice piled up at the foot of the volcano as much as twenty feet thick. More ash filled the sky and drifted northeast, eventually falling to the ground and blanketing most of the Pacific Northwest from present-day northern California up through British Columbia and as far east as Utah and Alberta. Some of that ash washed into creeks and rivers that flowed west into the Pacific Ocean, where the microscopic glass shards mingled with other river sediments and settled on the continental shelf and slope.

Mazama ash functioned almost like a geological Rosetta stone, a known touchpoint in the geological calendar of the Pacific Northwest. Its presence in these cores meant that every turbidite above the Mazama ash had been triggered in the years since about 5677 B.C., and everything below it had been triggered before the ashfall.

Goldfinger wasn't surprised to see variation, from one core to another, in the thickness of the sediment layers. What was a little startling was the consistency in the number of layers he was counting. Thirteen distinct layers above the Mazama ash layer, thirteen distinct turbidites appearing in core after core. It didn't seem to matter if the core had been pulled up near the Canadian border or off the mouth of the Columbia or one hundred or two hundred miles to the south: every single core had those same thirteen turbidite layers above the Mazama ash. There were some additional layers in some of the cores collected most recently, as the cruise had headed south. But all of them had in common those same thirteen turbidites.

These weren't random slope failures or slides from geographically isolated events such as storms or even distant tsunamis, Goldfinger realized in a heady rush. Only one thing could make turbidites in all these places, all at the same time: the great shaking that would follow a rupture of the Cascadia Subduction Zone.

Adams, Goldfinger realized with a grin, had been right after all. *Amazing,* he found himself thinking, not for the first time, *how much information you can extract from a tube of mud.*

By the time the crew had filed into the lab, anyone looking at Goldfinger could see that something had changed. Not that the compact, fit, fortyish Goldfinger wasn't ordinarily upbeat; he almost always had a smile on his face. But his mood now was nothing short of ebullient. You set out to prove a theory wrong, he mused, and if your methods are solid, even if you're proven wrong, you've done something right. And sometimes, like today, something game-changing. This turbidite record might stretch thousands of years further into the past than the 3,500-year record onshore geologists had found in the coastal marshes, and the longer the record, the more scientists would know about recurrence intervals for Cascadia's great quakes. The more that was known about recurrence intervals, the more solid would be the predictions about when Northwesterners might expect the next great quake, which was of course what the public wanted to know. Turbidites might even reveal the relative magnitude of Cascadia's earthquakes. A decade of marsh stratigraphy at Willapa Bay and the Columbia River and the Necanicum River in Seaside and all the big and little bays on the Oregon coast and south all the way to Mad River Slough and Humboldt Bay in California had built an irrefutable case for Cascadia's seismicity, up to and including assigning a firm date to the last Big One—the event apparently recorded in the topmost turbidite in the cores Goldfinger was now admiring in the *Melville's* lab. If he was right, Goldfinger realized, he might be witnessing the start of a whole new field of inquiry, not just for Cascadia but for subduction zones around the world. If that was the case, he knew that the real work—and the fun—was only just beginning. More cores to pull, analysis that might take years to complete, and formulation of responses to all the naysayers out there—people just like himself, just minutes before.

All of which was going through his mind when he paused for questions from the crew. The *Melville's* chief engineer was the first to raise his hand.

"I've been on I-don't-know-how-many scientific cruises like this one," the engineer began, "and I've been to a lot of these show-and-tells where the scientists explain what their research is about, just like you did here. To tell you the truth, I never understand any of it."

At this, the engineer paused and smiled broadly, slowly nodding his approval to Goldfinger.

"But I can count to thirteen."

28

Star of the Sea

TOM, 1999

IT WAS A GLORIOUS SEPTEMBER AFTERNOON, the sky azure, the air at least 80 degrees outside St. Mary Star of the Sea, the Catholic church perched on the hillside above downtown Astoria. It was even hotter inside the church, with the low-angled sun baking the church's south wall and illuminating the stained-glass windows above the altar. But it wasn't just the heat that caused the best man to wilt, then slump, and finally collapse into the arms of an alarmed maid of honor as the wedding party followed Tom and Kirsten down the aisle, to the happy clamor of applause against the Handel recessional coming though the church's sound system. Low blood sugar was the culprit, and a change in medications, apparently. As for the bride and groom—he in black tux, she in white satin with beaded bodice, a white veil cascading halfway down her back—they were two paces ahead, smiling and striding down the aisle, oblivious to the little drama playing out at the front of the church. Even Pat's passing out couldn't mar the moment.

Tom had expected some drama anyway, with so much family together in one place: all four of Tom's siblings and their spouses and their collective nine children, plus Kirsten's father and three of her four siblings. Tom had thought it might be his older brother who would behave badly, given Chris's nature and their deteriorating relationship. But Chris kept it together, grudgingly wearing the tuxedo Kirsten had required of the ushers but planning to change back into his blue jeans for the reception at the community center in Seaside.

> Hail, Queen of heaven, the ocean star,
> Guide of the wanderer here below,
> Thrown on life's surge, we claim thy care,
> Save us from peril and from woe.

Tom had proposed to Kirsten the previous Valentine's Day. It was a cold Sunday, and they had been hanging out in the living room at Tom's house when he left the room for a moment, returned, soberly approached Kirsten—then busted out laughing. As hard as he tried, he couldn't stop thinking of the time Kirsten had painted eyes on either side of her chin and tied a kerchief over her dark hair, then posed upside down to be "interviewed" on video as "Chinagin," her upside-down mouth looking comically bucktoothed, her forehead Chinagin's chin, her chin his bald head. It was funny as hell at the time, and though a marriage proposal was clearly not an occasion to be laughing out loud, Tom just couldn't help himself. He knew it was nerves: he'd been thinking about this moment for months, gearing himself up, knowing that Kirsten wouldn't wait forever. But Kirsten as Chinagin: it *had* been hilarious. More to the point: Kirsten was the kind of woman who would forgive a man for chortling through a marriage proposal, just the kind of woman Tom wanted for a lifelong companion. "Okay," she'd said with a grin after he finally got the words out. He had offered Kirsten his mother's engagement ring, from her marriage to his father: a simple .35-carat diamond solitaire. Sized for Bobbie's slim finger, the ring was a bit tight on Kirsten's hand. But it served its purpose: a promise, tangible proof of their intentions until they could go ring shopping together and Kirsten could find a ring her size, and more her style.

St. Mary, just blocks from the college where Tom and Kirsten had met five years earlier, had been Kirsten's choice. She was raised in the church, and still attended services. Tom, a lapsed Episcopalian, had willingly submitted to the priest's scrutiny, agreeing to raise as Catholic any children that resulted from their union. It was an easy promise to make. Two months before the wedding, Kirsten had spent the better part of a week on Tom's living room couch, painfully recovering from emergency surgery of a kind that would rule out pregnancy. Besides, she was by this time forty-one, he forty-five. They could adopt, but Tom wasn't sure he wanted to start parenting this late in life. With nine nephews and nieces, there were already plenty of kids around to spoil. Between their work and other interests—art making for Kirsten, and a long list of volunteer commitments for Tom—their lives were already full, their motivation to add parenting to that mix low. Which was, in fact, another thing Tom appreciated about Kirsten: she never begrudged his many night meetings and the hours he spent holed up studying budgets and writing minutes.

Not until a couple of years after his return to Seaside in 1993 had Tom finally accepted the obvious: that his relationship with Pam in Spokane was going nowhere. They had never fought, not once. Whether that was the hallmark of a great relationship, Tom wasn't certain, but he knew he liked the ease of it. The problem, as Tom saw it, came down to an unusual quirk of Pam's: she lived in constant fear of catastrophe. A funnel cloud appearing out of the clear blue sky. A drunk driver crossing the center line at just the wrong moment. A sinkhole opening in the street, an earthquake raining bricks. Tom could only speculate about the cause of Pam's disaster fixation—he assumed it had something to do with her father's death in a car accident when she was still a girl—but he was intimately familiar with its effects. Pam didn't mind riding in a car with someone else at the wheel, but she herself didn't drive, would not ride a bicycle, was afraid of trains, did not like to fly. She was comfortable in Spokane, where she had a good job and could get around by bus. Tom understood that small-town Seaside, with its lack of public transit, would be challenging for her. But even before he moved back to Seaside, he knew that that wasn't the half of it. The disaster Pam feared most of all, the one that seemed to symbolize for her every unpredictable, random, rare, devastating cataclysm, was a tsunami. Until recently, tsunamis hadn't existed as much more than metaphors for Pam: the least likely of actual disasters. But now she was beginning to read in the newspapers about evidence of gigantic tsunamis in the distant past on the Pacific Northwest coast, and predictions of another one sometime in the future. *Are you ever going to be able to come down to Seaside?* Tom had finally asked her on the phone one day. *No*, she had said. Simple as that. *I was afraid of that*, he'd said. There really had been nothing left to say. He had considered moving back to Spokane, but by the mid-1990s, there weren't many job opportunities for professional prospectors in the inland Northwest. And he knew the tsunami wasn't all there was to it. Pam could have, in that same phone call, begged him to move back to Spokane, or even asked him if he would consider it. But she hadn't. In fact, she hadn't even mentioned it as a possibility. *Well, I'll talk to you later*, was what Tom had finally said before hanging up.

Tom had met Kirsten Huling in the Extended Learning Office at Clatsop Community College in summer 1994, when he first inquired about teaching there; she served as the college's liaison with adjunct professors such as he, had helped him navigate the college's particular bureaucracy. It was a pleasure to run into someone like her in an office like that: she was quietly competent, efficient but not officious, not flirty but friendly, and quick with a smile. She had what Tom thought of as a kind of exotic Polynesian beauty, though Polynesian was one ethnicity she couldn't claim: her maternal grandparents were Japanese immigrants, her father mostly Irish and Dutch mingled with Mexican and Native American from the Southwest. When, in small talk with Tom a year or so into his teaching gig, Kirsten had mentioned that she was headed to southern California to fetch some furniture from her family's home but wasn't sure her car would survive the trip, Tom had done what he considered the obvious thing: he had offered her the use of his own pickup truck. When, at the end of that school year, he had hosted an end-of-class picnic on the beach near his house for his geology students, it seemed natural to extend the invitation to Kirsten too.

When, a month later, the whole Horning clan—siblings and cousins, their spouses and kids—gathered at the end of Twenty-Sixth Avenue to watch the Fourth of July fireworks and set off a few of their own, Kirsten had joined them at Tom's suggestion. There were kids everywhere that night, waving sparklers in the dark, hovering at the edge of the big beach bonfire. Kirsten had noticed one little boy, a cousin's son, younger than the others, playing by himself, seemingly overwhelmed by the night's noises and tumult. Apparently Tom noticed too. As they were all leaving the beach, Kirsten watched as Tom reached down and scooped up the youngster, carrying the now-smiling child all the way back to the house. Tom, the extrovert, keeper of the family home and traditions, was also the guy who noticed the child lost at the margins and who unceremoniously salvaged the boy's holiday. Witnessing that gesture, Kirsten had felt something melt inside. But, in fact, she had already fallen for him. It must have been obvious in her face—to everyone, apparently, but Tom himself. "Do you see the way she's looking at you?" Judy had quietly asked Tom earlier in the evening, elbowing him and gesturing at Kirsten. He had to admit he hadn't; apparently his skill at reading women was on a par with his ability to read green rocks. But now he noticed. Within weeks, they were a couple. Within months, their relationship had an air of inevitability. Within a year or two, Tom's nephews were starting to needle Tom at family gatherings, pleading, "When are you and Kirsten going to get married?"

Born in Hawaii, raised mostly in Hollywood, Kirsten had moved to the Oregon coast in 1991 when her sister Mara, a newly commissioned ensign in the Coast Guard, had been assigned to the cutter *Resolute* and had urged

Kirsten to move with her to Astoria, the vessel's home port, to keep her company. Kirsten had been a German major in college, had taught for a couple of years in Austria, but since moving back to Los Angeles had held a series of unrewarding jobs. She was single; nothing really held her there. Astoria seemed to offer the opportunity to start fresh, to take the art classes she had always wanted to take, to remake herself as an artist. It hadn't quite worked out that way. But she liked working at the college and took full advantage of the opportunity she had as an employee to take art classes for free. She lived in a circa-1925 house she had bought on Grand Street, a bungalow with a magnificent view of the Columbia River and a bottomless list of needed repairs. From her bedroom, she could hear the melodious barking of the sea lions in the East Mooring Basin and, too many nights, the raucous laughter of drunks stumbling up the hillside stairs from the Desdemona Club on Marine Drive. Moving into Tom's house in Seaside would stretch her work commute from five minutes to twenty-five. But it was an easy trade to make. She had already grown fond of the view above the kitchen sink, of the estuary and salt marsh and spruce forest at Neawanna Point.

And Kirsten wasn't bothered by the prospect of earthquake or tsunami; growing up in Hawaii and California, she had already survived at least one of each. The same wave from Alaska that had nearly washed the Horning cottage out to sea in 1964 had lapped Hawaii too, though all Kirsten could recall of it was her glee at hearing that school had been cancelled. Seven years later, she was in southern California, still asleep in the bunk bed she shared with her sister, when the magnitude 6.6 Sylmar earthquake struck early the morning of February 9, 1971. The epicenter was just twenty miles from their home in Hollywood. No one in her household was hurt, though Kirsten could vividly recall her father rushing into her bedroom, staggering like a drunk across the buckling floor.

Seven years earlier almost to the day, Kirsten's mother had been holding her own against heart disease and had nearly finished treatment for breast cancer when she was felled by a seemingly innocuous virus, probably something she picked up from one of the kindergarteners she taught. Kirsten knew a thing or two about the uncertainty of life, and she had made a certain peace with it. What she knew: she was happy with Tom, happy at the prospect of living in that house with that view and that congenial a companion. She wasn't about to squander it by worrying about a natural disaster that might or might not occur in her lifetime.

The Source of life, of grace, of love,
Homage we pay on bended knee:
Do thou, bright Queen, Star of the Sea,
Pray for thy children, pray for me.

29

Penrose

TOM, 2000

THE BALLROOM AT THE BEST WESTERN Ocean View Resort still smelled, to Tom's nose, a little sickly sweet, like a cat with a bladder infection: out-gassing from the four-by-eight sheets of pressed-wood particleboard Tom had purchased and hauled to the hotel and leaned against three of the ballroom's walls. All but the east wall, where tall curtains now blocked the view of the forested hills and the sky, dour this cloudy day in June. Every scientist at the Geological Society of America's Great Cascadia Earthquake Tricentennial Penrose Conference seemed to have brought a poster to display on those boards. Tom had spent most of the lunch break hopping from group to group clustered at the posters, happy just to listen—latest findings, new hypotheses, controversial conclusions—and occasionally throw in his own two cents' worth. There was an exhilaratingly democratic tone to the conference: Tom—a professional geologist, well educated but unpublished, the local expert—rubbing shoulders and swapping ideas with academic luminaries, chief scientists, leading geological consultants, civil engineers. And a few small-town fire chiefs and big-city emergency managers. Conveners of the four-day conference chose to focus Day One on seismic hazards and hazard mitigation before diving into the science, and they had invited local emergency managers and other public officials to take part.

Tom was actually more than a participant at the conference—and less. Truth was, he hadn't actually paid the conference fees. He figured his attendance was payback for having volunteered to help plan the thing. It was Tom who had proposed the Ocean View, on the Prom in Seaside, after Brian Atwater had called asking for his suggestions. Atwater's requirements: a conference facility for up to one hundred participants somewhere on the Cascadia coast, preferably on the beach. And it had to be close enough to Willapa Bay to allow Atwater to offer a day trip, on Day Two, piling sixty people into rented canoes and floating them past his study site on the morning's ebb tide. Tom had also scouted the road portion of Atwater's field trip, identifying Bruceport County Park as the lunch spot. Atwater had wanted a place where scientists could display sediment cores they had pulled from the continental slope, seashore marshes, and a lake on the southern Oregon coast. Now Tom was glad he had picked that site with its big, log-walled, shake-roofed picnic shelter: the forecast was for rain, followed by showers.

The list of conference participants read like a who's-who of Cascadia researchers. Some Tom knew well: Curt Peterson, of course, and his protégés Mark Darienzo and Brooke Fiedorowicz. And Chris Goldfinger from Oregon State: Tom had spent a couple of weeks at sea with Chris back in September 1997, shortly before Goldfinger had turned to turbidites. A member of the scientific crew had dropped out at the last minute, and Tom had taken his place, spending a long day folded inside a two-man submarine smaller than a Honda Civic and puttering around the base of a mud volcano two miles deep off Ocean Shores, Washington. George Priest and the gang from DOGAMI were there. Some attendees—from Japan, Italy, England, Chile, and throughout the United States and Canada—Tom knew only by their research. Gary Carver, for example, down from Kodiak, Alaska, and Alan Nelson from Denver. And Kenji Satake, without whom there might not have been a tricentennial date to commemorate. George Plafker, still with the USGS in Menlo Park, was the only participant at the conference who had also attended the very first Penrose conference at Asilomar, back in 1969, when the theory of plate tectonics was still being assembled. Tom had been especially eager to seek out Plafker, dean of the 1964 Alaska Good Friday earthquake researchers, to regale him with stories of that event's impact here in Seaside. And there were young geologists Tom hadn't yet met, people like Rob Witter, a consultant out of Walnut Creek, California, who'd done work on the southern Oregon coast.

That first day's morning session hadn't held many surprises for Tom. It was planned as a kind of Earthquakes 101 course, mostly for the benefit of the non-scientists in attendance, covering all the hazards that come with a very big earthquake: the ground shaking that knocks things down, on the coast and for one hundred or more miles inland; the tsunamis that follow, devastating shoreline communities; the subsidence that leaves the coast lowered, bays enlarged, and beaches flooded even after the tsunami is over; the liquefaction that cracks roads; the ubiquitous landslides. The afternoon was dedicated to speakers addressing mitigation of those hazards, such as the chief structural engineer for the City of Portland, who bemoaned the poor condition of the state's buildings. Because Oregonians had not been aware of the seismic history of their state until quite recently, the state's building code didn't adequately address it. That left perhaps 90 percent of the buildings in the state "exceptionally vulnerable" to earthquakes.

That would be a big problem for people farther inland, but would be the least of Seaside's problems when the Big One struck. Which is why Tom straightened in his chair when Nobuo Shuto, a professor of policy studies at Iwate Prefectural University in Japan, approached the podium.

Shuto—square shouldered, square jawed, with salt-and-pepper hair— had focused his research on itemizing and enumerating tsunami disasters: Human lives lost to drowning, or to injuries from floating debris, or to disease following exposure to too much seawater or, worse, fuel oil or gasoline floating in the tsunami's outflow. Houses washed off their foundations or collapsed on site or standing but saturated. Crumbled seawalls. Destruction of bridges and highways, or their closure due to erosion or debris, and ports unusable due to debris and colliding fishing boats. Water supplies cut off, electrical and telephone lines washed away. Fishing industries devastated by damaged and lost boats and gear and by destruction of marinas. Industry and commerce halted when goods were wrecked and offices flooded. Crops destroyed, and farmers' fields buried in sand. Oceanfront forests damaged by incursion of seawater and trees ripped out by their roots and subsequent soil erosion. Oil spills, floating fires, electrical fires—for, as counterintuitive as it may seem, Shuto explained, fire is one of the big hazards of tsunamis.

It was a long list, sobering. Shuto had studied the effects of several tsunamis in Southeast Asia and had quantified the kinds of damage one could expect from tsunamis of varying heights—from a three-foot tsunami that a stout wooden building might withstand to a twenty-six-foot wave that could destroy even stone buildings and tear up reinforced concrete structures.

How to prepare? You can build defensive structures, Shuto said— seawalls and breakwaters at bay mouths and barrier forests—but these work only against small tsunamis. You can alter your city planning, limiting construction of homes and schools and police and fire stations and hospitals to high ground and requiring shoreline structures to be built of reinforced concrete; a tsunami might blow out the doors and windows of such buildings, he said, but the remaining structure can block floating debris and help shield more vulnerable wooden structures inland.

But it was defensive *systems*—development of evacuation systems and of what he termed a "disaster culture"—that were the "last and best method to save human lives," Shuto said. Tom found himself leaning forward, not wanting to miss a word, straining a bit to understand Shuto's accented English and slightly strained syntax. "Shaking caused by an earthquake is a natural tsunami warning," Shuto intoned. "Public education is indispensible to make forgettable human beings to continue the precious former experiences and to prepare for the next tsunami."

Shuto had also studied death rates in tsunamis and correlated them to water depth, plotting them on a graph. Now he flashed that graph on the screen at the front of the darkened ballroom. Tiny circles and squares and Xs and triangles were scattered between two bold parallel lines that swept up and to the right, much like a steep, cresting wave. The graph's legend was written in *kanji*, but with Shuto's explanation, it wasn't hard to get the gist. The symbols represented varieties of death from tsunamis. Points on the y-axis were numbered from 0.01 to 100.00: the percentage of the population dead from one tsunami cause or another. The x-axis was numbered 1 to 20: tsunami heights in meters. An arrow pointed at the number 2, with the message "At 2 m, loss of life begins." Approaching the ten-meter line—thirty-three feet—Shuto-san's graph indicated a death rate of 90 to 95 percent.

Mothers and fathers, children, grandparents, tourists, farmers, taxi drivers, city councilors, waitresses, business owners. This was not a computer model of projected fatalities but a tabulation of individual deaths from actual tsunamis. People. As Tom stared at the screen, visions of Seaside crowded his mind's eye: the weathered shingles and crisp white trim and cobble fences of the beachfront houses, the clammers out digging early in the morning on minus tides, the tourists parked on the beach or bumping elbows on Broadway, the kids pouring out of the middle school on a weekday afternoon, headed home or to the skateboard park next door. The math was easy. At a tsunami depth of fifteen feet, half the town's residents would be gone, according to Shuto's figures. At thirty feet—which was possible; the geologic record indicated tsunamis had reached that depth and deeper here in the past—something like 90 percent of the town would die if they failed to evacuate. Some fifty-five hundred people filled Seaside alone, and that would be on a winter weekday. On a summer day, when the beaches and hotels and downtown streets were packed with tourists, that number could be three times higher.

Shuto clicked, and a new image appeared on the screen: a grainy black-and-white photograph, hard to make out at first. As Tom stared, he realized it was a huge wall of water, splashing against something and filled with debris. What kind of debris, he couldn't quite make out; it just looked messy. It was, Shuto explained, a rare photo of an actual tsunami. Because tsunamis happen quickly, without warning beyond the earthquake that can precede it by mere minutes, and because those close enough to snap a picture don't tend to survive, few photos of them existed.

The photo shouldn't have surprised Tom, but it did. He knew tsunamis pick up debris, had seen it himself as a child, the sharp angles of household goods and whole trees streaming out the mouth of the bay, silhouetted by car headlights and the lights of the houses at Little Beach. But seeing the photo somehow brought it home to him: it's not so much the water that kills in a tsunami; it's all the stuff *in* the water. In his mind, he was watching boulders rolling underwater, cars bobbing and bumping on the surface, uprooted Sitka spruce tumbling past second-story windows, those windows blowing out, the buildings themselves breaking loose.

This—Shuto's presentation—was what his neighbors in Seaside needed to see and hear and prepare for, Tom realized. This is what we, or our children or their grandchildren, will someday face. Water, and a whole lot more. Death, unless we're able to get ourselves over a river and a creek and up a hill.

A big, deadly mess.

30

Sumatra

SHERWOOD, OREGON, 2004

Knowledge, even when it is exact, does not often lead to appropriate actions because we tend to forget what we know, or forget how to process it properly if we do not pay attention, even when we are experts.

—Nassim Nicholas Taleb, *The Black Swan:*
The Impact of the Highly Improbable

IN 2004, GEORGE PRIEST OFFICIALLY RETIRED from his job with Oregon's Department of Geology and Mineral Industries, though he was still working part-time for the agency. The timing seemed to have worked out almost perfectly, not just for himself but for DOGAMI.

The years since he finished drawing what was still being called, inelegantly, the "SB 379 line" had been every bit as busy as those that had led up to it. SB 379 was a statutory line, a guide for new construction, nothing more. It didn't actually outlaw anything; it was only as good as the local building inspectors charged with enforcing it. Meanwhile science, and DOGAMI, had marched on. In 1996 the agency had received funding to refine its earthquake and tsunami modeling methods, based on new information from geologists, and to produce a more detailed and, scientists believed, realistic inundation map for Yaquina Bay at Newport. This map had an entirely different purpose from the SB 379 maps. It was not a guide for new construction but an evacuation map based on a spectrum of possible Cascadia tsunamis: it showed not where to build but where and how high to run. DOGAMI had mined additional funding sources and eventually produced such maps for the state's twenty-seven main coastal population centers. Oregon Emergency Management then jumped in to design and distribute evacuation brochures based on DOGAMI's new maps.

Seaside officials had been less than enthusiastic when Priest and Lou Clark, DOGAMI's outreach coordinator, had arrived in Seaside in July 1998 to present their new tsunami evacuation maps for the city. What city manager wants to hear that virtually his entire town can expect to someday be wiped off the map in a big tsunami? But within six months, city officials had begun to shed their denial and had started wrapping their minds around the new normal: a catastrophic threat with a vague timeframe—say, sometime in the next few hundred years. Even if buildings are destroyed by the tsunami, they came to understand, people can survive if they have bridges to cross, high ground to run to, and time enough to do it.

As it happened, voters in Seaside had just agreed to establish the Greater Seaside Urban Renewal District to fund a slate of improvements throughout the town. Reconstruction of the city's older bridges—seismic upgrades of a kind that should keep them standing through a magnitude 9 earthquake—had been languishing toward the bottom of a long menu of possible expenditures. Now the city's priorities shifted. Within three years, four of the city's twelve bridges had been rebuilt, bridges bristling with rebar, their pilings buried deep underground. Rebuilding the oldest bridge in town—the 1924 Broadway Bridge, crossing the Necanicum in the heart of the downtown tourist strip—would be a herculean task, prohibitively expensive, or so went the thinking. Instead, the First Avenue Bridge just one block away was replaced, and one year later, the Broadway East bridge over Neawanna Creek, giving beachgoers, schoolchildren, merchants and their customers along Broadway, and everyone else an evacuation route that shot nearly straight east, toward the safety of the Sunset Hills neighborhood. At the same time, both Twelfth Avenue bridges—over the Necanicum River and over Neawanna Creek—were replaced, providing an escape route for kids at the high school and neighborhoods on the north side of town. The new bridges might or might not weather a fifty-foot tsunami. The point was, they almost certainly could withstand the shaking of the earthquake that would precede it, intact enough for pedestrians if not for car traffic. With four bridges likely to remain standing, at least some people in Seaside now had a chance.

As soon as the third of the four bridges was completed, the city began posting tsunami evacuation signs throughout town, with arrows showing in what direction to run, and by the next year, the Oregon Highway Department was posting warning signs wherever U.S. 101 descended into the tsunami inundation zone. The signs featured a white stick figure dashing uphill against a blue background, with a caricatured wave nipping at his heels. The new signage was greeted with mixed reviews. In Cannon Beach, Japanese tourists could be seen posing next to the signs, mugging for the camera. Some signs disappeared overnight, rumored to have been snatched as souvenirs by surfers or, in some cases, by locals concerned about the dampening effect such signage might have on coastal economic development. T-shirt makers embraced the logo, tweaked it slightly, and displayed it on Beefy Ts with such taglines as "In Case of Tsunami, Grab Beer and Run."

All the larger communities on the Oregon coast now had tsunami evacuation maps. There was always more to do—Priest's replacement, who hadn't yet been hired, wouldn't lack for projects—but the geologist felt good about what he had accomplished. He had begun his career at DOGAMI before anyone had a clue that the Cascadia Subduction Zone was a thing to worry about. By the time he had retired, the entire Oregon coast had been mapped—in some cases, twice—for tsunami inundation and evacuation, and towns and cities were at least starting to think about how to prepare for the disaster geologists predicted was coming sometime in the next few hundred years. He'd moved into semi-retirement a couple of months earlier but, so far, had managed to stick to the part-time schedule he'd set for himself, splitting his time between DOGAMI's Newport and Portland offices as he had for many years, living much of the time in the small cabin he and his wife had bought in Newport but maintaining their family home outside of Portland.

Which is where he was on Christmas Day 2004: at home in Sherwood with his wife and grown stepchildren, celebrating the holiday. They had long since finished opening presents and were starting to think about dinner when he first heard the news on TV: a massive tsunami had followed an earthquake in the Indian Ocean.

He promptly logged on to his home computer to check the quake's magnitude: early estimates suggested it might be as high as magnitude 9. If that was true, Priest knew, that would make it one of the biggest in recorded history. As far as he knew, there is only one way to make an earthquake of that size: with a subduction zone, which was almost necessarily underwater. A subduction earthquake of that size would be followed by a proportionately large tsunami. Immediately he knew a tragedy of massive proportions was in the making.

Of course he was right. Crude video footage of the tsunami inundating beach resorts in Thailand and roaring up river valleys at Banda Aceh,

Indonesia, was all over the television news, footage the likes of which even he—a geologist charged with helping prepare the Oregon coast for that very kind of event—had never seen. What came to be called the 2004 Sumatra-Andaman earthquake and tsunami damaged or destroyed cities, villages, and resort communities throughout South Asia and East Africa, killing some 227,900 people, including thousands who went missing but whose bodies were never found, and displaced another 1.7 million. At a magnitude later set at 9.1, it was the largest earthquake in the world since the 1964 Alaska quake, in an area whose coastlines were vastly more densely populated than those of Alaska, then or now.

As soon as preliminary casualty figures began coming out, Priest knew he was witnessing a game-changer. Not only was it among the deadliest natural disasters in human history, but this earthquake, which occurred where the India Plate is subducting beneath the Burma Plate, was certain to spark new interest in tsunami preparedness on the Oregon coast. The City of Cannon Beach had been looking for help to improve its tsunami evacuation maps; DOGAMI, shy of funds, had not yet provided the small town with updated maps, leaving it to rely on what Al Aya and Paul Visher had created back in 1993. Up to now, Priest could provide Cannon Beach only with generalizations—rough approximations of the run-up of tsunamis in the past, extrapolated from marsh coring done elsewhere on the Oregon coast. To get more specific was going to take more money, money that had been hard to come by from politicians who couldn't seem to understand what he and other geologists were talking about. How do you explain a mega-tsunami to policymakers or anyone else who had never seen one? How do you explain the overwhelming numbers of people it would kill, the destruction of entire cities in a matter of minutes?

Now Priest didn't have to. Sumatra had done it for him. If the Cascadia Subduction Zone had not been on American politicians' radars before this, he figured, it would be now.

31

Higher Ground

SEASIDE, OREGON, 2008

*Suddenly, earthquake science stopped being fun, and as a scientist,
I began to feel like the watchman on the castle walls warning
about barbarians at the gate, begging people to take me seriously.*
—Robert S. Yeats, *Living With Earthquakes in the Pacific Northwest*

THE FORECAST WAS FOR POSSIBLE SHOWERS, nothing more than that, and the clouds were high and hardly threatening. So Doug Dougherty grabbed his parka and a large rolled map, stopped at his secretary's desk to let her know his plans, and headed out to his blue Honda SUV parked outside the Seaside School District administration building. The Honda was uncluttered, and recently washed: neat, like superintendent Dougherty himself, in his khakis and freshly ironed cotton shirt, his waterproof parka mud-free. Only his hiking boots testified to his recent ramblings. He headed north on U.S. 101, then just past the high school turned east onto Crown Camp Road, which cut a twisting route up through clear-cuts stubbled with fresh stumps and stands of tall second-growth forest. He slowed as he passed through what was left of Crown Camp—the white, single-story headquarters building designed by John Horning in the 1950s that now housed an outpost of the timber giant Weyerhaeuser, owner of about one quarter of the land in Clatsop County.

From Crown Camp, Dougherty steered the Honda onto the Necanicum Mainline, a gravel road pocked with dark potholes half-filled with rainwater and littered with the previous season's alder leaves. He continued west and south until he could see, through the fringe of trees between the road and a yawning clear-cut, the ocean beach two miles to the west and several hundred feet down. He slowed, then pulled over at a wide spot, leaving plenty of room for passing log trucks. Grabbing the map and his daypack, he got out, locked up, and began walking west, descending through the clear-cut, threading his way among the Douglas fir stumps.

Dougherty was looking for—well, he wasn't exactly sure what he was looking for as he picked his way down the slope, staying high to avoid having to climb out of creek-cut ravines. But it would have to be pretty level—a flat piece of ground, someplace without a lot of clay soil that might weaken in a quake. A large, flat area not too far from town, but high up in the hills—at least one hundred feet in elevation. A place too high for the highest possible tsunami to reach, on solid ground highly unlikely to slide in the event of a massive earthquake.

That's why he had driven up into the forestlands east of Seaside: there was no place within the city limits that fit this description, nothing. The land he sought would need to be easily accessible from the center of Seaside, however. Size would matter; he was going to need a lot of ground. Dougherty was no geologist, no engineer—all of his academic degrees, up to and including his PhD, were in education—but he'd grown up on Oregon's Mount Hood, had always been comfortable in the outdoors, felt he had a sense for the land. And even if he didn't quite know what he was looking for, he thought he might recognize it when he saw it, or at least recognize when he'd found something worth asking a geologist to investigate.

Besides, it soothed him, between seemingly endless meetings and phone calls and e-mails, to just be out looking, hiking cross-country alone on a weekday afternoon like this or even on a weekend with his family. Doing nothing was simply not an option, and talking about doing something but not actually *doing* it was even worse in Dougherty's book, given geologists' recent estimates of the size of tsunamis that have hit this coast in the past ten thousand years, their predictions that something similar is certain to strike the coast here sometime in the foreseeable future, quite possibly within this or the next generation of Seaside School District students.

Hopefully not today.

DOUG DOUGHERTY HAD BEEN WORRYING about the tsunami since long before he became superintendent. Safety had always been Dougherty's number one priority—almost something of an obsession. Student achievement was important, of course, but the district's kids were doing great, with standardized test scores now on a par with the best schools in the state. And frankly, what would a student's SAT scores be worth if a tsunami came along and washed her out to sea? Given what Dougherty had been hearing lately from coastal geologists, it wasn't a far-fetched scenario. He had spent his entire career at the Seaside School District, first as a PE teacher, and then, after detours to earn master's and doctorate degrees, becoming the district's director of curriculum and instruction. In 1992, with student achievement on an upward trend and the budget tight, the superintendent had asked him to take on the principal's job at Cannon Beach Elementary School as well.

As principal jobs went, this one was a piece of cake. Cannon Beach School sat on what was probably the prettiest location for a school in the state of Oregon, maybe in the whole country. The school sat just a block from the beach, on the south bank of Ecola Creek. Kids on the swings could glimpse the ocean waves if they pumped their legs hard enough.

Cannon Beach, population fifteen hundred, had long ago evolved from a logging town to an artist colony, and then—as is often the fate of artist colonies—a gentrified tourist mecca. Not all the kids' families were well off—most didn't own the art galleries and bistros and motels but, rather, worked in them—but were, as a rule, keenly interested in their children's education. They volunteered in classrooms and they showed up at Parent Teacher Organization meetings. Dougherty split his time between the school and the district office—a six-mile drive over Tillamook Head. Especially during recess, when Dougherty sometimes pulled playground duty, with gulls keening and kids grabbing at his legs or begging for a boost onto the monkey bars, it was hard to imagine a better place to work.

But for all the satisfaction the job gave Dougherty, it also came with a growing sense of unease. He had been following the progress of proposed legislation dealing with the recently discovered risk of major earthquakes and tsunamis in western Oregon. Dougherty, ever alert to safety issues, began to educate himself on the topic. Since the Monmouth meeting in 1987, DOGAMI had been urging legislators to take seriously the state's emerging seismic risk. The eventual result was a pair of senate bills passed in 1995. One created a task force to recommend changes to the building code addressing seismic retrofitting of existing buildings: earthquake readiness. The other dealt with tsunami risk. It would require new hospitals, fire stations, conference facilities, schools, and other "critical and essential facilities" to be built above a tsunami inundation zone that would be defined by state geologists.

Dougherty knew that, regardless of how high the line was drawn, Cannon Beach Elementary School would be in the zone. For evidence, there was the school's old swing set. Two of its steel uprights had been kinked by a drift log that crashed into it when the tsunami from Alaska sent water surging over the playground in 1964: mute testimony to the school's vulnerability. That same tsunami had also swept away the bridge spanning Ecola Creek, at the edge of the schoolyard. That bridge had later been rebuilt to the same engineering standards to which it had originally been built: in other words, demonstrably not tsunami-proof. New building codes could keep a school district from building a new school on the beach, but they would do nothing to keep kids in Cannon Beach's existing grade school safe.

He began meeting with his teaching staff for some sobering discussions. Few had yet heard of the Cascadia Subduction Zone; few understood that the next tsunami, the Big One, wouldn't come from someplace on the other side of the Pacific Rim. It would come from a fault zone right off Oregon's shore. If it struck on a school day, everyone—from kindergarteners to custodian— would have less than a half hour to exit the school and get to high ground.

He also collaborated with City Hall and the fire department to hold a series of town hall meetings, to introduce the term Cascadia Subduction Zone and explain what it meant, to talk about the difference between a distant tsunami—the only kind to have visited the Oregon coast in living memory— and a local tsunami, the kind geologists were now warning about. It was challenging, striking the right note: somewhere east of panic but west of denial. It's not like this is likely to happen tomorrow, he and the fire chief stressed at those meetings. "It's the type of thing that happens maybe once every five to six hundred years," Dougherty explained, adding that scientists figured the last one happened around three hundred years ago. "When it *does* happen, here's what you want to be thinking about." Have a plan, he said. Know the quickest route from your house to the hills. Have a bag of emergency supplies packed and sitting next to the door.

It took a while for the information to sink in: Dougherty could see it in their eyes. The vacant stares of boredom. The twinkles of amusement, humoring the over-eager school principal. Even scorn: Chicken Little, insisting the sky was falling. In fact, Dougherty noticed, people seemed to need to hear the message three or four times before they got it, before they realized that this was different from the tsunami warnings of the past. He could recognize the moment the message sunk in: *A massive tsunami is going to wipe out this town, possibly in your lifetime. You'd best be ready.*

"Don't be surprised to see kids and teachers walking across the bridge, holding hands and heading up the hill," he told parents and other citizens gathered at the town hall meetings, because by this time he and the teachers had already begun drilling the students on how to evacuate the school in a hurry. The route he'd scouted led through a gate in the cyclone fence at the corner of the playground and across the bridge over Ecola Creek. The bridge might be cracked from the earthquake, engineers had told him, but it should still be standing and able to take the weight of a line of children on foot. From the creek, the kids followed the road's edge to U.S. 101, well above sea level in its ascent of Tillamook Head. His destination was the fifty-foot elevation line—plenty high, according to the experts at DOGAMI.

The first drills had been pretty chaotic: the very first day he'd made the kids repeat the whole routine three times. But it didn't take long before they had it down. That was Dougherty's goal: if he did his job right, no one should have to tell the kids what to do when the time came. Then came an unexpected pop quiz: a sonic boom, rattling windows and nerves in Cannon Beach and leading the fire district to trigger the citywide tsunami warning alarm on a day when Dougherty and all the teachers happened to be at an out-of-town workshop. As he later heard it, the children themselves—kindergarteners included—led the startled substitute teachers out of the building, over the bridge, and up the hill.

Somehow a Portland television station got wind of the tsunami drills in Cannon Beach, and then another, and pretty soon, *National Geographic* showed up to interview Dougherty and to film the kids dropping in place, covering their heads, and holding onto desks, then filing out of their classrooms and across Ecola Creek, hand in hand. It was an irresistible story: cuteness, emerging science, and mortal danger served up on the iconically scenic Oregon coast. The sudden celebrity was a little heady for a small-town school principal. One star turn seemed to lead to another, and more media requests followed. The kids loved it and quickly become pros, cheerfully indifferent, no longer hamming at the first appearance of a camera.

Dougherty agreed to work with DOGAMI on a preparedness video, then he helped the kids produce a thirteen-minute film of their own that was distributed to every school district on the Oregon coast. "It takes our students less than two minutes to evacuate the building and cross the bridge," Dougherty boasts in the students' documentary. "In less than fifteen minutes, our students are atop of Highway 101."

It was a comforting thought—but was it true? What if the bridge should fail? What if someone were injured by falling debris during the earthquake?

In fact, even before *National Geographic* arrived, before the tsunami drills had become institutionalized at Cannon Beach School, Dougherty knew there was only one way he could be confident he had done all he could to ensure the safety of the students.

He had to move the school.

Funding issues aside—and they would be formidable, in a town with some of the highest property values in the state—the question was, where? Not only was land expensive, but most of it was within the tsunami inundation zone George Priest had drawn for the state building code in 1995. Then Dougherty learned about a seventeen-acre tract of undeveloped land south of downtown that the city had set aside in the 1970s as a future school site. It seemed ideal; by chance, its elevation was between about thirty and forty-two feet, putting it just above the inundation zone. But interest in the site by the school district piqued the interest of citizens with other agendas, who succeeded in convincing the city council to rezone the property exclusively for open space. Undeterred, Dougherty engaged the help of Cannon Beach's city planner, and together they launched a search for alternative sites. One promising piece of ground turned out to be the type likely to liquefy in an earthquake. Another was prone to landslides.

By this time, Dougherty had relinquished his split role as curriculum director and school principal to become superintendent of the entire district. Now he was responsible for not one school but five, but he also had a bigger bully pulpit from which to preach his message of tsunami safety. He had moved into his new office the same month in 1998 that George Priest delivered the first-ever tsunami inundation map to Seaside. Almost the entire city was color-coded yellow, for "evacuate from this area." The state had not yet drawn a similar map for Cannon Beach, but it was obvious that map would be a very yellow map as well.

How high was high enough? That question ate at Dougherty, through the years of searching for alternative school sites in Cannon Beach, through the thrice-yearly tsunami drills that, by the late 1990s, Dougherty had institutionalized not just at Cannon Beach but at the district's other three low-lying schools as well. What was so magical about the SB 379 line? Dougherty wondered. Could he be confident that schools above that line would be safe? Was it possible that a tsunami in certain locations—Cannon Beach, for instance—could *exceed* that height?

It was indeed possible, George Priest replied when Dougherty finally put the question to him in 2004. The line Priest had drawn was intended only as a reasonable building code guideline—an estimate of the height of an average, most-likely local tsunami, based on the best geological evidence available at the time of past local tsunamis in selected sites along the Oregon coast. It was never intended as a worst-case-scenario line: the height to which a citizen should climb in order to be certain he or she was above the reach of an incoming tsunami.

That line, Dougherty realized, was the line he wanted: the worst-case line. Not the average, most likely height, but the elevation of demonstrated safety, the line of least concern, the height that would guarantee Superintendent Dougherty a full night's sleep. That was where he wanted his schools to be. And the logical place to start was at Cannon Beach Elementary, a school full of young children, located at the mouth of a creek, in a community that was already tsunami-savvy.

"So can you tell me exactly what has happened specifically in Cannon Beach, in the geological past?" he asked Priest. I don't know if you'll need to dig core samples or what, Dougherty told the geologist. He simply wanted to know exactly how high local tsunamis of the past had risen in Cannon Beach. Specifically.

"Yes, we can look at that," Priest responded. Cannon Beach had actually been next on Priest's own mental to-do list for tsunami evacuation mapping. He'd been a bit chagrined that DOGAMI had produced evacuation maps for twenty-seven other coastal communities but hadn't yet done Cannon Beach. And Cannon Beach struck Priest as the ideal spot to systematically test current tsunami models against the actual tsunami inundation that could be inferred from distribution of tsunami sand deposits known to be in the marshes around Ecola Creek. DOGAMI was in the process of hiring a young geologist, thirty-eight-year-old Rob Witter, to take over from Priest and to allow him, eventually, to retire, and the agency had some federal funding to pay for some computer modeling. But it would take a couple of months to get the job done and require additional funds to cover Witter's time—$30,000. Dougherty took that figure to the Cannon Beach City Council and fire department, each of which agreed to contribute $10,000 to the project, matching $10,000 from the school district.

It was 2006 by the time the study got under way. Curt Peterson and a colleague at Portland State University had already begun investigating the Cannon Beach marshes and the surrounding hills for tsunami deposits. Witter took the lead for DOGAMI, collaborating with the PSU researchers.

Within a couple of weeks, Dougherty got a phone call from Priest. "Hey, Doug," Priest began, the geologist's calm, deliberate speaking style by now a familiar sound to Dougherty. "This is going to be a much larger project than we'd originally planned. I'm afraid we're not going to be able to finish this in just a couple of months like we talked about. It might take us another year or two to do this right." Priest and Witter had managed to convince higher-ups at DOGAMI of the need to do more than produce the standard type of inundation map they'd been doing but, this time, to do a thorough study of Cascadia tsunamis that would take much longer. Geologists drilling core samples on the continental slope off the Oregon coast, Priest explained to Dougherty, had found evidence of at least twenty huge earthquakes in the past ten thousand years, some of them even larger than magnitude 9. The 2004 Sumatra-Andaman tsunami had also broadened geologists' thinking about what was possible; now they knew just what a tsunami from a magnitude 9.1 earthquake could look like.

Coincidentally a new funding source had also appeared: a result of the 2004 Sumatra tsunami. That disaster on the other side of the globe, captured on video for the world to watch, had for the first time spurred Congress to take seriously the tsunami threat to the United States' own coastline—a threat concentrated at the country's northwest corner. Late in 2006 President George W. Bush signed the U.S. Tsunami Warning and Education Act, authorizing funding to "strengthen the tsunami detection, forecast, warning and mitigation program of the National Oceanic and Atmospheric Administration." Now DOGAMI had the funds it needed not only for detailed field studies but for computer modeling vastly more sophisticated than had been possible a decade earlier.

SOME DAYS, AS GEOLOGISTS WERE PLUNGING corers into the marshes behind Cannon Beach, Dougherty wondered if he shouldn't just close the town's school, lock it up at the end of that school day. Declare that the risk to student lives was too high and that henceforth the students would be bused to Seaside Heights Elementary School. Seaside Heights had opened in 1973, replacing old Central School downtown, and only by pure luck did the district choose a site that was seventy feet above sea level. Cannon Beach Elementary was tiny—just one small class for each grade—and

Seaside Heights could easily absorb the additional students and teachers. But that kind of precipitous action, taken with the best intentions, would rile parents—*involved* parents, people whose support he would need to get a new school built.

And what about the kids in Gearhart? Gearhart Elementary was another old school, a year older than the one in Cannon Beach, and it sat in a dune swale just inches above sea level. The only high point kids from that school could reach in a reasonable time frame was the forty-five-foot-tall dune directly west of the school, which geologists suggested would be safe only in an average-sized Cascadia tsunami, nothing bigger.

The risk to the high school students was as bad as or worse than that to kids in Cannon Beach—and there were a lot more students at the high school. The school sat directly east of the mouth of the Necanicum River in Seaside. Knowing what Dougherty knew now about the way tsunamis get magnified in river and bay mouths, that was not a good thing. Students would have, at most, twenty or thirty minutes to run a mile and a quarter down U.S. 101 to the cineplex, then east up Twelfth Avenue across a bridge (recently rebuilt to current seismic standards) to reach a safe elevation. The athletes would make it; everyone else would be iffy. The middle school wasn't quite as vulnerable; it sat on Broadway, between the river and the creek, and the bridge that crossed the creek was also new. Assuming it withstood the quake, and it should, the kids had to jog only a half-mile or so to reach what should be a safe height. That meant most of the middle schoolers would make it. Some of the older teachers and the disabled kids would not. None of the schools, however, had been built to withstand an earthquake—any earthquake, much less one that shook for three to seven minutes.

And then there were the Seaside School District administrative offices, where Dougherty himself worked, housed in the old hospital building on the west bank of the Necanicum River. Its ground floor had flooded during the 1964 tsunami—a mere *distant* tsunami. In the event of a big Cascadia earthquake, the district staff would have one river and one fat creek to cross, on bridges that had not yet been rebuilt and weren't likely to be standing after the earthquake.

Six buildings, five of them located not even fifteen feet above mean sea level. It was a dubious distinction: no school district in the United States had more buildings located in a tsunami inundation zone. If the earthquake and tsunami happened today, most of the students and staff in the Seaside School District would not survive.

Then came the meeting where Dougherty got his first glimpse of how idiosyncratic science can be, how colored the results of a research project can be by who happens to be in the room on a given day. DOGAMI had been holding regular meetings with stakeholders and scientists as the Cannon Beach study progressed. The question this day was where to draw the new evacuation line. Previous evacuation maps had drawn the line somewhat higher than the SB 379 building code line. But the current study had lain bare the full spectrum of possibilities. The question of the day: how extreme an event should be used as the basis for the evacuation line: the average run-up height, or the height that would put you above the majority of tsunamis, or the height above which no tsunami has ever reached?

Dougherty listened to the debate for a while. Then he spoke up. "Well, thinking as a parent," he began, "I would want to know what the highest level was that a tsunami has ever reached in Cannon Beach. That's where I would want the school my children attended to be located. Would I be comfortable if that school was at the 60 percent confidence level for tsunami inundation?" he asked rhetorically. "No."

He looked around at the assembled scientists, who were nodding now. It was as if they'd forgotten that all these numbers were more than data that could be manipulated in a variety of ways with amazing speed on high-powered computers that could generate a range of scenarios. What Dougherty saw when he looked at the numbers: the last girl in the line of kids heading up the hillside during the last tsunami drill in Cannon Beach, her green leggings and her polka-dotted sneakers and her T-shirt with the glittery logo, chatting distractedly alongside a teacher who was doing her best to hurry the girl up the hillside, a girl who had been going up the hill just as fast as she could go, a speed that Dougherty knew would never be quite fast enough.

THE HIGH SCHOOL CAFETORIUM WAS ALREADY FULL by 8 a.m. the morning of May 7, 2008, not with students but with staff: teachers, aides, custodians, lunch ladies, school secretaries, everyone from the district office and all five schools. The timing of this in-service day was fortuitous; it gave Doug Dougherty an opportunity to share his news with the district's entire staff before it hit the newspapers.

"As you've heard me say many times, student safety is the highest priority," he began. "Nothing we do, or don't do, should ever compromise their safety.

"I think most of you understand that we can have two types of tsunamis here on the coast." He launched into a brief review of the geology, then reiterated the district's attempts to find a safer site for Cannon Beach Elementary, the thrice-yearly tsunami drills the district now required at all four low-lying schools, and the project geologists had begun in 2006 to more precisely gauge the height tsunamis had reached in Cannon Beach in the past. Two years later, that project—DOGAMI's top priority statewide—was still under way.

"That brings us to today," Dougherty said, smiling the smile he smiled when he was nervous or about to deliver weighty news. The results of the Cannon Beach study, he said, were surprising the scientists. "The report is still in draft form," Dougherty said, warming to his key message, "but just yesterday I learned that the *Oregonian* newspaper has formally petitioned the governor to release the data in its current state. They're planning to run a front-page story in Sunday's paper."

At that, Dougherty dimmed the lights and switched on a projector, flashing a map of Cannon Beach on the screen behind him. Lines in bright blue, red, orange, yellow, and pale green ran from north to south in squiggly patterns—geologists' latest estimates of the run-up of Cascadia tsunamis over the past several thousand years.

The lowest line—blue—scribed a ragged circle around the heart of downtown, the school grounds, and the low-lying neighborhood north of the creek. Virtually no houses stood above the worst-case, the highest, green line: the height geologists were estimating a tsunami had run up some fourteen hundred years earlier.

"What the scientists from DOGAMI and others have found is that if we want to build a tsunami-safe school for the children in Cannon Beach, we can't build it in Cannon Beach. There is literally no suitable property within the city of Cannon Beach to safely relocate the school. Nothing."

And if there were no tsunami-safe sites to relocate the school in Cannon Beach, he said, chances are the project would tell a similar story in Seaside and Gearhart.

DOGAMI also happened to be wrapping up a statewide seismic needs assessment of schools and emergency facilities throughout Oregon: 3,352 buildings, judged for their ability to withstand the biggest earthquake likely to occur at that location within a "reasonable period of time." Seaside Heights was judged to be in good shape. But most of the remaining school buildings in the district were rated at high risk of collapse.

The room was quiet. Dougherty waited a few seconds to let the information settle in.

"What we're looking at now," he continued, "is relocating all the schools in the district to a single campus in the hills east of town," he said. Not rebuilding individual schools at separate sites in the three communities, but moving all the students except those at Seaside Heights to one big campus. Two elementary schools, the middle school, the high school, administrative offices, playing fields, bus barns—everything. "We're starting to look at forestland above town," Dougherty went on. "That's going to require pushing out the urban growth boundary, because there's nothing high enough inside the city limits of any of our towns. So that's what we're looking at now: moving all the schools to one site."

He scanned the crowd; a few people were exchanging glances, but no one seemed particularly alarmed. The superintendent planned to build a whole new school campus? That was an ambitious proposition.

On the other hand, given what they had already known, and what they knew now, how could he not?

32

Aftershock

CAHUIL, CHILE, 2010

*Never had I witnessed anything like this before, though I had
heard of earthquakes. I found myself rocking on my horse and I
moved to and fro with him like a child in a cradle, expecting the
ground to open at any moment and reveal an abyss to engulf me
and all around me.*

—John James Audubon

NOT UNTIL THE AFTERNOON of their first day on the coast did Rob Witter
realize he and Laurie would actually be doing all the driving in Chile. "I
can drive," Leonardo had assured his companions as they filled out the
paperwork at the rental car agency in Santiago that morning. And he
could—as long as the car was an automatic. Which the little Toyota Yaris
they had rented wasn't. Rob, a forty-two-year-old father of three young
daughters, had felt like the dad after Leonardo climbed behind the wheel at
Pichilemu, talking Leonardo through the curves in the road to Cahuil: "OK,
now let out the clutch, *easy* ..." The next five days were going to be long
enough without Leonardo grinding the gears at every change of grade; no
need to add driving lessons to the itinerary. And Leonardo had something
more important to contribute than driving. Given Rob's and Laurie's limited
Spanish, Leonardo's translation skills would be essential.

Witter, jetlagged and low on sleep less than twenty-four hours after his
arrival, was still surprised to find himself in Chile. Not nearly as surprised
as the Chilean people had been sixteen days earlier—Feb. 27, 2010—when
a magnitude 8.8 earthquake had shattered the pre-dawn stillness and
generated a tsunami that swept the beaches and tore through port towns
along three hundred miles of Chile's central coast. Some 521 people had
been killed, most of them by the quake itself; quake-savvy Chileans had
generally known to run uphill. This would be the first of two trips to Chile
for Rob in 2010, as it turned out. He had already been planning a trip in

May to Viña del Mar, where he was to give a talk sponsored by the USGS on the work he was doing in Oregon, updating the work his predecessor, George Priest, had started more than fifteen years earlier, using more recent fieldwork and more sophisticated computer modeling to determine the theoretical upper limit of run-up from a local tsunami along that state's coastline: basically, re-drawing the evacuation line people should run toward after an earthquake. Then the Chilean quake struck, and he had been invited to join an ad hoc group of scientists and other quake specialists for some immediate post-quake fieldwork. Members of the Geo-Engineering Extreme Events Reconnaissance team had assembled the night before at an office on Magdalena Street in downtown Santiago. That's where Rob had first met the other members of his sub-group: Leonardo Dorador, a grad student in the Geophysics Institute at the Universidad de Chile, and Laurie Johnson, a risk management consultant from San Francisco. Their assignment: spend five days on the country's south-central coast surveying land-level changes— uplift and subsidence—caused by the tectonic forces that had also triggered the magnitude 8.8 quake.

They had gotten an early start the next morning, renting the car and stocking up on supplies and groceries in Santiago. It had been slightly more than two weeks since the earthquake and tsunami, and the coast was still in disarray. Many of the businesses not destroyed by the quake and tsunami were now closed, and few hotels were operating. Martial law was in effect in the larger coastal cities; unless they could find lodging and check in before the 9 p.m. curfew, they would need to be out of town and on their way back over the mountains on broken roads at the end of the day in order to find a place to sleep—then go back the next day. The drive to the coast that first morning—about twice as far as the distance from Portland to Cannon Beach or Seaside, Rob judged—took hours, between slowdowns for road repairs and detours along alternate routes and over temporary bridges. *And this in a country engineered to deal with quakes*, he found himself thinking. *Oregon will not fare so well.* They reached the coast at Pichilemu, the northernmost of the Chilean towns severely damaged by the tsunami. Jet lag notwithstanding, Rob had been eager to get to the coast and get to work. He had spent the previous seventeen-plus years of his life studying the causes and effects of large earthquakes and tsunamis, most of them hundreds or thousands of years in the past. Never before had he had the opportunity to see the effects of such an event this fresh.

Entering Pichilemu, Rob had steered the Yaris to a gas station to fill up and to quiz the employees about their recollections of the pre-quake landscape and the run-up of the tsunami. Actually Leonardo quizzed them while Rob and Laurie listened and picked up what they could, waiting for Leonardo's translation.

It was there, standing outside the gas station in Pichilemu, where Rob felt his first aftershock—a small one, but enough to freeze him and Laurie in place while the ground simmered and the two of them looked around, wide eyed.

"Un temblor," the station attendant had said, nodding his head and looking from Rob to Laurie, smiling.

Oh, right, Rob had thought, laughing out loud at himself. *We're in an* earthquake *area!*

His driving skills notwithstanding, Leonardo proved to be an ideal traveling companion. He was good-natured and energetic and eager to help out, willingly posing—for scale—in Rob's documentary photographs: Leonardo, smiling, in jeans and striped polo shirt, standing with one arm upraised next to a tree with tsunami-borne marsh grasses strewn on branches higher than his reach. Laurie, who ran her own risk management firm, had high cheek bones and a ponytail, with a ball cap often parked on her head. Unlike Rob, Laurie had lots of firsthand experience with quakes. Judging from the mild look of panic she adopted at every aftershock, those quakes seemed to have left some scars.

Rob himself was of medium height, with a trim physique and a ready smile and a handsome face not quite obscured by rectangular black-rimmed eyeglasses—think Clark Kent with a slightly receding hairline. He even had a superpower, one that allowed him to look deep into the earth and see, with startling clarity, the future. It was called paleoseismology.

Not until a few months into graduate school had he begun to fully develop this power. Rob had grown up in the Seattle suburbs and had been a biology major at Whitman College in central Washington. But by senior year his passion for the life sciences had waned. On a whim, he signed up for a geology class and was immediately smitten by plate tectonics: the high drama of continental and oceanic crusts colliding and piling up over eons and right this very minute. At the University of Oregon, where he went for grad school, he met a visiting professor who was studying geological evidence of huge earthquakes and tsunamis on the southern Oregon coast. Such evidence had already been found at several other sites on the Pacific Northwest coast, Rob learned, but the mouths of the Coquille and Sixes rivers and adjacent coastal plain were virgin territory. His plan had been to complete a master's degree at UO, but after he and his supervisor—geologist Harvey Kelsey, Tom Stephens' old boss in Humboldt County—successfully applied for funding from the National Science Foundation, a doctoral dissertation was born. After graduating, he spent nearly six years in the San Francisco Bay area working for William Lettis and Associates, a leading geotechnical consulting firm launched by a former student of Gary Carver.

The job posting in *GSA Today* had appeared at just the right time. Oregon's Department of Geology and Mineral Industries was seeking a coastal geologist with a background in earthquake and tsunami hazards to live and work on the coast. It sounded, to Witter, like a perfect fit. He and his wife had about lost patience with the Bay Area traffic. Lettis and Associates was poised to merge with a larger consulting company, which would mean he'd be sent to work on bigger projects, farther from home, for longer periods of time—things like geoscience hazard studies for nuclear power plant sites on the East Coast. By then they had two young daughters, and Witter was interested in spending more, not less, time at home. Newport, Oregon, was an appealing next step. It was a busy town in the middle of the Oregon coast with plenty of other well-educated, outdoorsy types like himself around, many of them employed at Hatfield Marine Science Center, across Yaquina Bay.

Witter's office was on the second floor of a small, slightly cramped office building in what passed for downtown in Newport: a collection of forgettable government and commercial buildings clumped on the bluff north of the bay. The office ambience didn't really matter to Witter. When he was there, he was usually at a computer, modeling tsunami run-up and earthquake slip. But that first year, 2006, he was often not there at all. The Cannon Beach job had blown up into something much bigger than anyone had expected and was expanding into a project to remap the entire Oregon coast for tsunami evacuation, based on new research about the worst-case run-up. After Cannon Beach finally wrapped up in 2008, Witter's next target area had been Bandon, on the southern coast. There were good reasons for going there next. For one thing, it was on the south coast, which had issues distinct from those on the north coast. And a lot of fieldwork had already been done there, some of it by Witter himself. Now there would be new maps, better maps: printed and interactive online maps. And the funding wasn't just for fieldwork and computer time and mapping. It was also providing a few months of employment for community organizers on the coast to rally local emergency responders and community leaders to start their own preparedness efforts. Tsunamis anywhere in the world, Witter had found, are typically followed by an uptick in funding for tsunami preparation and mitigation. But it never lasts long. One of these days that big tsunami wouldn't be somewhere else. It would be on the Pacific Northwest coast. It would be nice if people at least had the information to know what to do, where to run, when that day came.

THE TEAM HAD SLEPT THAT FIRST NIGHT at the coast in accommodations they found in Cahuil following a busy afternoon collecting data at Pichilemu and Laguna Petrel, a small estuary on the northern edge of town. Here the coastline appeared to have subsided a bit during the earthquake—less than a foot—and the tsunami had banged up the first row of oceanfront houses. Otherwise the damage wasn't extreme. The night in Cahuil would be the last night they would spend on the coast itself. The next day they drove south to Bucalemu, a riverside hamlet that, to Rob, looked like a miniature Waldport, Oregon, its town center arrayed on the south side of a wide river mouth. Here the tsunami damage had been worse, with many more damaged buildings and destroyed campgrounds and a disappeared beach. Just *gone.* A foot and a half of subsidence had apparently drowned this sandy beach where residents used to launch their fishing boats. Now there was just a severely eroded, wave-beaten rocky shoreline—eroded either by the tsunami itself, or by wave action following subsidence, or both.

The data collecting in Bucalemu took up most of the morning, followed by a stop to see the impact the earthquake had had on some lakes at Laguna Torca and a look at a damaged bridge at Lipimavida, where a tsunami surge at least twenty-seven feet high had churned up a river valley and pushed a bus nearly one hundred yards upstream. So it was early evening by the time they approached Constitución, a city they knew had been severely damaged by the quake and tsunami. Two-thirds of the casualties nationwide—three hundred fifty people—had occurred in Constitución. That number might have been much higher had the city not coincidentally held a tsunami drill just two weeks earlier.

Rob and the others knew Constitución would be a mess; they had seen the images back home on TV. People they met on the road from Cahuil had warned them that it was bad: "Constitución doesn't exist," was how a government worker in Bucalemu had put it. But driving into town and seeing it for himself, Witter quickly learned, was completely different from hearing it described or even seeing it on the news.

The devastation was simply overwhelming. Reading through his journal months later, Rob would find his entry for Constitución oddly devoid of detail; he had apparently been stunned into silence. Big fishing boats rested on their keels on city streets, marooned where the tsunami had dumped them. Where buildings had once stood, Witter now saw heaps of wood and twisted metal. Front-end loaders were driving down the city streets, still clearing debris more than two weeks after the event. Had the tsunami drill saved any lives? Had it caused one or two people to make, in a split-second, the right decision: to *not* turn back and grab a coat, a photo album, a cat before heading out the door in the dark? To *not* jump into the car and, instead, to head to high ground on foot? To scramble out of bed and tug at a parent to get out, get going, or to not even wait for mom or dad and to just take off running uphill alone?

Two weeks ago, people here had been going about their business or enjoying a few days of vacation, just like the day before, Witter reflected, peering out through the car's windows as they crept down the cracked, wet streets. A thought came to Rob, kept tugging at him as he scanned this modest coastal city built on a low plain, a river running through the middle of town, tourist hotels lining the beachfront, a city now in shambles.

That's Seaside, Oregon, he kept thinking. *Exactly.*

THE TEAM HADN'T LINGERED IN CONSTITUCIÓN. There was the curfew to think about; they needed to be out of town and on their way to that night's lodging at Salto del Laja before 9 p.m. And they'd begun to feel like disaster tourists: just another car clogging the roads, macabre sightseers ogling the devastation and evidence of shattered lives. Given the chaos and the curfew, they wouldn't be able to do any data collection anyway.

Rob recalled the twinge of guilt he'd felt back in Oregon about stepping away from the mapping project for this trip to Chile; the little state agency he worked for, DOGAMI, was overwhelmed trying to complete a project that had grown increasingly complex and time-consuming, and his absence for a week meant another week's delay in getting it done. So it had seemed a bit luxurious to leave work for a junket in South America. Now, after all Rob had seen—the striking analogs to a future post-quake Oregon coast—the trip didn't feel frivolous at all.

But looking back months later, it wasn't Constitución and the devastation there and elsewhere that Witter found himself returning to again and again in his thoughts. It was a much more personal memory, formed that first night on the coast at Cahuil—an intimate and unexpectedly sublime encounter.

It had been late afternoon by the time the team had finally rolled into Cahuil, the little destination beach town nine miles south of Pichilemu. Or *former* destination. Normally Cahuil would still have been crammed with kiteboarders in early March, the end of the Chilean summer. But the town was practically deserted. There were no vacationers. None. And most locals had apparently fled inland to stay with relatives. The team had worked the rest of that afternoon and early evening in Cahuil, measuring and mapping, examining the damaged highway bridge, the campground swept clean by the tsunami, the funky vacation huts the tsunami had toppled and washed upstream. Cahuil sat on a small coastal river widening to a lagoon near its mouth, reminding Rob of southern Oregon's Coquille River, where he'd spent many days in the field, first in grad school, later on contract with Lettis and Associates, most recently doing marsh coring for the mapping project. It was nearly dark by the time they returned to a cluster of vacation cabins painted in pastel colors that they had spotted earlier in the evening on a bluff north of the river. Leonardo, brandishing his cell phone, managed to track down the owner, who was thrilled to have customers.

The *casas rusticas* were located on what Rob recognized as an uplifted marine terrace, above the reach of the tsunami. They weren't all that rustic, and they were apparently undamaged by the quake: no cracks in the wallboard, no problems with the plumbing. *Just like the earthquake literature says*, Rob noted. Subduction zone earthquakes tend to be hell on bridges and tall buildings, but small wood-frame houses like these generally just flex with it and settle back down. Not until 10 or 10:30 p.m. did the team finally gather in one of the three cabañas they had rented to grill some *queso y jamón* sandwiches and cut up some veggies and open a bottle of Chilean red they'd brought from Santiago.

And there it was again, a little *frisson* in the floor, a shimmering. It must have been shaking like this on and off all day, Rob realized, but the team had spent much of the day in the car, and he hadn't felt the aftershocks then. Laurie froze in place, her expression grim, her hands clutching the seat of her chair and waiting for the shaking to pass. Leonardo took it in stride, opening a *cerveza* and throwing more bread and cheese on the griddle. Rob, a student of earthquakes who until today had experienced but one tiny temblor in his whole life, was enchanted.

"Man, this is amazing!" he exclaimed. "It's going through my whole body!" He began timing the quakes, noting the seconds between the P-wave hitting—a sharp jolt, like train cars collapsing together at a stop—and the S-wave, followed by surface waves like the rolling of the ocean, trailing off like distant thunder underfoot. Time seemed to slow as he waited for the aftershocks he now knew to expect: the wrench, then the shake, then the rumbling to a close.

Before that day, the physics of earthquake waves had existed as nothing more than words and diagrams on a page for Rob. He knew that the P-wave, the primary wave, is fast and thus usually the first thing a seismograph picks up; it travels by compressing and expanding particles in the earth, like a Slinky toy, pushing and pulling, moving as easily through liquid and gas as through solid rock. The S-wave follows—S for secondary, or sideways, because that's how it moves, in a side-to-side or up-and-down motion, but only through rock; it can't pass through liquid or gas. Or S for *slow*: S-waves typically arrive a split-second or several seconds behind the P-wave. Then come the lingering surface waves, the ground waves, moving across the earth like riffles on a lake. The pattern of waves can be clearly seen on a seismogram: two bursts of short, sharp peaks with a little fluttering in between, followed by ground waves' higher, bigger, sharper peaks diminishing to a slightly wavering line representing what seismologists call *background noise*: the Earth at rest.

That night, tucked between the sheets in the cabaña at Cahuil, Rob barely slept. He seemed to doze, dreamily, his body a human seismograph recording every movement of the Earth. A P-wave would hit, like someone had punched the bed, *pop!*, rousing him from sleep, and he would wait in the stretched-out timeframe of semi-consciousness before the S-wave struck, passing through him, and the surface waves began to rock him like a baby in a cradle, a nestling in a tall tree, a boat on the ocean. Sometimes that first punch was more of a nudge, a slight shove, before a shiver of quaking started up. After the sharper P-waves, the rolling would last longer, trailing out into the gentlest-possible, nearly imperceptible surface waves, like resting his hand on a purring cat, or a sleeping baby, or a lover, a movement almost sensed rather than felt.

It was like riding a train rumbling across an endless plain, he mused, half-conscious in the half-light of dawn. But it wasn't a train. It was the Earth. And never had Rob felt so in touch with the Earth, so coupled to it.

33

S–XXL

Gold Beach, Oregon, 2011

What can seem secure to us at all, if the globe itself is shaken?
If the only thing that is immovable and firm in the world, that
carries everything, moves backwards and forwards, if the Earth
has lost her characteristic, the stand still—where should we find a
holding in our fear, a refuge?

—Seneca

"Hello? Testing. Can you hear me now? Is this on? OK! Good evening, everybody, thank you for coming to the first Gold Beach tsunami preparedness meeting—the first of many!"

It was three minutes past 6 p.m., the advertised start time, and people were still streaming into the cavernous Curry County Fairground Events Center in Gold Beach, at the mouth of the Rogue River on the southern Oregon coast. Most were gray-haired, in jeans and sweatshirts and winter parkas, a few in billed caps—they had ducked in out of a dark, wet January evening. There were younger adults as well, some of them trailing children, and volunteers were scurrying to add to the hundred and fifty chairs that already were nearly filled. Along one side of the room were tables stacked with brochures and staffed by representatives of the Red Cross, the Forest Service, the National Weather Service, the Community Emergency Response Team, and ham radio hobbyists. Chief of Police P. J. Janik stood nearby, surveying the crowd, ramrod straight in his dress blues. A fiddler and keyboard player were jamming in one corner: the Pistol River Trio, minus the bass player. A smattering of snacks was laid out on banquet tables at the back of the room: plates of cut-up vegetables and dip, trays of crackers, a yellow cake with chocolate frosting covered with plastic wrap, with a sign: "For Intermission."

Technically it was not the first tsunami preparedness meeting in Gold Beach, "Tsunami Deb" Sterling whispered to Rob Witter, the two of them standing off to one side at the front of the room. One year earlier, Oregon Emergency Management had done something similar, but only ten people had shown up. Sterling credited this night's burgeoning crowd to better advertising and to advance work by herself and others, including the community organizer that Oregon's Department of Geology and Mineral Industries had recently hired for the southern Oregon coast. Tonight's meeting was part of a three-night road show DOGAMI had organized to introduce communities on the southern coast to the new tsunami evacuation maps Witter and his team had drafted and to help pump up local tsunami preparedness efforts.

"*Thankyouverymuch!*" James Roddy began, channeling Elvis. "Thank you for coming out on a *beautiful* evening! Is it raining outside? Yeah, I thought so. I thought the southern Oregon coast was supposed to be nice and dry all the time!

"Before we get started," he continued, mock-serious, "Rob and I are both state employees, and it's after six o'clock, so we're going to need to take a break in about fifteen minutes"—chuckling from the audience—"and then probably another one about every fifteen minutes …" Gales of laughter now; he'd broken the ice.

"This is a federally funded program—none of your state tax dollars are going toward this"—more chuckling—"and it's a four-year program to help coastal communities in Oregon get better prepared for what could happen in the future. And there's new research that indicates that this big earthquake that we talk about and the tsunami that it generates is much more likely than we ever thought possible even just a few years ago."

Roddy was the earth science officer for DOGAMI. His background was in television; he knew how to engage a crowd, how to balance the silly with the serious. He launched into a basic explanation of plate tectonics for an audience that, for the most part, would have finished high school back when the theory of continental drift was nothing but a fairy tale. It was a canned talk, but he gave it a local spin: eleven Oregon schools were located in the tsunami inundation zone, as DOGAMI had recently redefined it, and Gold Beach's elementary and high schools were two of them.

"If you don't think this can happen in the United States, I have one word for you: Katrina," Roddy said, invoking the 2005 hurricane that had wreaked havoc across a wide swath of coastal Louisiana and Mississippi including, famously, New Orleans. "Who knew a modern city in this country could be devastated by a natural disaster? A million people displaced, more than thirteen hundred people dead. In parts of the Gulf Coast, help didn't arrive for two or three *weeks*. And if you think the federal government is going to do a better job here in the Pacific Northwest"—more chuckles from the audience—"with more people affected, good for you!"

He started to wind down, but not before (cue soundtrack of Mongolian throat singers) telling a story about how the sea gypsies of the Andaman Islands had survived the 2004 Indian Ocean tsunami because their legends, handed down generation to generation, instructed them in how to behave after the Earth shakes; and then another story about elephants spontaneously heading to the hills in Thailand in advance of that same tsunami; and then a tongue-in-cheek suggestion that elephants be imported to the southern Oregon coast; and then a film clip of elephants riding surfboards accompanied by Henry Mancini's tuneful "Baby Elephant Walk" from the 1962 film *Hatari!* By now the crowd was cracking up. Then, in an instant, Roddy sobered the mood, flashing photos of Chile, February 2010, not quite one year earlier: upside-down cars littering a roadway next to a collapsed freeway overpass; a red-and-white fishing boat marooned in a wet, debris-strewn city street.

"Disasters are inevitable," he wrapped up. "They're personal. It's the survival part that's optional."

"Thank, James Roddy—doesn't he do a great job?" said Sterling, joining the applause as Roddy handed her the microphone. Sterling was DOGAMI's coastwide tsunami outreach coordinator; she worked for Roddy from an office in Seaside. She asked for the lights to come back up and took a few minutes to give shout-outs to the local ham operators and CERT volunteers and first responders in attendance. Then, "Without further ado," she said, "I'd like to introduce Dr. Rob Witter!"

Witter, blinking in the spotlights, stepped up. This was his third presentation in as many nights on the southern Oregon coast, and he was still figuring out how to simplify the message, sharing the science without insulting people's intelligence or talking over their heads. No elephants on surfboards in his PowerPoint, but he did have an animation that he hoped would open some eyes.

"My job is to use the best available science to help you prepare for a worst-case event," Witter began, "and I'm going to show you a summary of the science that goes into making these new maps.

"Because we have not had historical tsunamis here in Oregon, we have to make the best educated guesses to develop an accurate tsunami evacuation map. We use science to do that. These tsunami evacuation maps I'm going to show you tonight represent the worst-case scenario. That is a very low-probability event. I hope that the next tsunami, which is inevitable, is not this big. But as long as you're heading uphill toward the tsunami evacuation zone, you're heading in the right direction.

"So a quick primer on the subduction earthquake cycle: James …?" He looked to Roddy, who cued the first slide. "Big earthquakes make big tsunamis. But we don't know how big the next one is going to be, and we don't know when it's going to occur. Next slide please?

"As the Juan de Fuca Plate subducts beneath North America—and that's happening right now, beneath our feet, these two plates are converging about one and a half inches per year—the friction between the two plates causes them to lock together, and it drags down the tip of the fault, and it pushes the coast up. When the stresses get so high that they're stronger than the strength of the fault, the fault will break, and the motions will be reversed. The coast goes down, and the seafloor lurches westward and raises the overlaying column of water, and that's what produces the tsunami. And it's this process that we're trying to model."

Witter referred to comments Roddy had made earlier in his presentation about the work of Chris Goldfinger at Oregon State University, who studied underwater landslides—turbidites—triggered by earthquakes in the Cascadia Subduction Zone. "As a result of his work, we now believe that more earthquakes have happened than we thought before. In fact, about twice as many." Forty of them in the past ten thousand years, he said: nineteen of those being complete ruptures of the entire six-hundred-mile-long plate boundary, an average of one about every five hundred years. "But they don't occur regularly spaced out. If they did, then we'd know how much time we have. Scientists worldwide have tried to predict earthquakes, but we have not been successful.

"The turbidite record also suggests that four of those forty earthquakes were of only three-quarters the length of the rupture zone, and there were about seventeen more that appear to be just limited to the southern part of Oregon. So that raises the hazard down here on the south coast. But since they happen more often, they're likely to be smaller, and the tsunamis are likely to be smaller too.

"I like to think of earthquake and tsunami events in terms of T-shirt size," he continued. This gimmick was Witter's latest attempt to make the science accessible, the message memorable. The last event, in A.D. 1700, he said, was an M. Some are S, some are L, "and there's a little data that suggests that there have even been giant extra-large and even extra-extra large quakes here," he said. Small-size quakes might be generated by just, say, thirty-three feet of slip—meaning, North America moves, in a matter of seconds, toward and on top of the Juan de Fuca Plate thirty-three feet. "In the worst-case scenario we're modeling, North America lurches seaward over the Juan de Fuca Plate more than 144 feet. In magnitude ranges, a small would be magnitude 8.7—and that's not small! That's the size of what happened last February in Chile—and up to a 9.1, which is equivalent to the 2004 Sumatra-Andaman Islands earthquake."

Quakes of that size would produce tsunamis of proportionate size, he said. "In Gold Beach, for the small tsunamis, we predict they'll reach run-ups of about forty feet high, on average." A more likely height for a tsunami at Gold Beach, he said, is sixty feet. "Large tsunamis may reach as high as eighty feet, and in the case of the extra-large and the extra-extra-large, we're looking at run-ups approaching one hundred and twenty feet.

"This is about double of what we thought previously."

He paused to change slides, and the room was silent. Not a cough. Not a shuffling foot.

"Why are the tsunamis that we predict now for southern Oregon so much bigger?" he asked rhetorically. "This is something that Sheriff Bishop wanted me to share with you, and this is important information. There are several reasons why we're predicting higher run-ups here in southern Oregon.

"For one thing, before the Sumatra tsunami, no one thought tsunamis could reach heights of one hundred feet or more. And no one had thought to model, on the Oregon coast, a tsunami that would result from a magnitude 9.5 quake of the kind that occurred in Chile in 1960.

"There are also some peculiarities to our coastline here." The continental shelf isn't nearly as wide in southern Oregon as it is in northern Oregon, he said. The underwater edge of the continent is roughly the edge of the Cascadia Subduction Zone, where one tectonic plate is attempting to override the other. Witter pointed to a map on the screen that illustrated his point clearly: a squiggly line in the ocean not quite parallel to the Oregon shoreline, a line about eighty miles west of Seaside but, in its slant to the southeast, less than forty miles west of the shoreline at Gold Beach. "This wide continental shelf"—gesturing to Washington and northern Oregon—"slows tsunamis down. Tsunamis shoal and lose their energy when they encounter the bottom of the ocean, and it reduces their size. The tsunamis are able to approach the coast with higher energy down here in southern Oregon. There's also the Rogue Channel and some other bathymetric features off the coast here that appear to be causing the wave"—computer-modeled—"to be very high."

A new map flashed on the screen, and anyone could see it was the shoreline at Gold Beach and the adjacent ocean, brilliantly colored. The warm colors, yellows and reds, indicated wave peaks, Witter explained, and the cool colors, the blues and greens, were the troughs of the waves. "So what you'll see is that this peak is uplifted by the deformation of the seafloor during the earthquake. That peak is going to move onshore."

Witter pointed to a time clock in a corner of the screen. "That's the time from when the earthquake happens. Go ahead, let 'er rip, James."

Roddy pushed a button on the laptop to start the computer animation. While Witter narrated, the audience watched the red-crested wave march toward their town with chilling inevitability. At the mouth of the Rogue River, the wave splintered into more waves that began to collide chaotically, then all of it—the colliding waves at the river mouth and the long, straight waves approaching the beach south of the river—collapsed onto the town. The time from the moment the earthquake started the clock ticking to the wave's arrival at the Gold Beach shoreline: less than fourteen minutes.

"It's important to remember that the first wave is not always the highest wave, and it's certainly not the last wave. These tsunamis will continue to arrive for hours and hours. In the event of an earthquake, the earthquake is your natural warning to head to high ground, and you should stay at that site for a long time. Don't return to the beach. Don't go back to low ground, because more waves will come. OK, next slide.

"Now let's talk about *distant* tsunamis. What's the warning for distant tsunamis?"

"Sirens," several voices warble from the audience.

"Sirens is one. And if you hear the siren, relax. Turn on your NOAA weather radio, turn on your TV, find out what's happening. 'Cause you have a lot of time. *Hours.* The closest big earthquake source we have for a distant tsunami is Alaska, and it takes four hours for that tsunami to reach Oregon."

The new tsunami evacuation maps, Witter said, show two evacuation zones: the lower-elevation orange zone is for the worst *distant* tsunami scenario that computer modelers have been able to imagine: a thirty-foot run-up. Higher than the tsunami in '64, but still ten feet lower than the smallest imaginable *local* tsunami. And the yellow zone—well, he'd already talked about the heights a local tsunami could reach.

Now a map patchworked with orange and yellow and green appeared on the screen: the new official tsunami evacuation map for Gold Beach and nearby Nesika Beach, although poster-size print-outs of the same map were already mounted on the Events Center walls in several places. Hand-outs of the map had long since disappeared from display tables. "I'm sorry we didn't print enough," Witter said, "but they're available on the county website, so you can get these as PDF files and download them on your computer.

"I want to point out that the tsunami inundation zones are bigger—that's the big difference between the new maps and the old maps. We want people to move to ground that is even higher than before."

On this map, orange was thick all along the outer coast—including the airport—and into the mouth of the river, site of the harbor and Coast Guard station. Yellow blanketed downtown and reached into ravines rising to the east, where Gold Beach's newish elementary school sat, at an elevation an untrained eye would have assumed was safe. The old high school was so low in elevation, so close to the beach, it was practically in the orange zone. U.S. Highway 101 was entirely in the yellow zone on this map, except where it crossed the Rogue River. That bridge was built in 1931 and, according to state highway engineers, was one of forty bridges on U.S. 101 in Oregon that was expected to be extensively damaged—if not destroyed—by a magnitude 9.0 quake.

Witter continued pointing out features of the new map: its legend and key messages were printed in both English and Spanish, and beach access points were noted, as were presumably safe post-disaster assembly areas. The map's colors were even chosen to be visible to the color blind, he said.

The siren is a warning only for a distant tsunami, Witter reiterated. For a local tsunami, the earthquake *is* the warning. "So if you're warned by an earthquake, protect yourself by ducking under a table: drop, cover, and hold to protect yourself from falling objects. Unreinforced masonry buildings, bookcases, things that topple down: that's what hurts people. Earthquakes don't kill people; it's the things that fall down on top of them that hurt them. And once the shaking stops, you've got about ten or fifteen minutes to get to high ground. OK, before we go to questions …" but Tsunami Deb stepped in.

"How many of you have questions?" she asked. By now, nearly three hundred people—one-tenth of Gold Beach's population—had squeezed into the room, but not even a half-dozen people raised a hand. She directed them to take their questions straight to the "brain trust"—Witter and the other presenters—after the meeting. And she introduced Curry County Undersheriff Bob Rector, "who would like to say a few words."

Up stepped a man dressed not in uniform but, like most everyone else, in jeans and an outdoor jacket. "I just want to touch on a few things and reinforce some things these guys have talked about so eloquently here tonight," the undersheriff began. "These guys have done a good job of laying out the new tsunami standards that we have to get used to. The threat, the hazard, is bigger than we thought before. Where we're at now is, we're really at the beginning as to how we process and deal with this new information. It's not rocket science, but the new map really gives us the ability to do things better than we could do before. Now, there's three points I want to reinforce.

"You need to plan to evacuate on foot. A lot of people have equipment and stores in their vehicle, and they plan to use that vehicle to get out of harm's way. The earthquake of the magnitude we're discussing here tonight is very liable to make the roads impassable. So plan to evacuate on foot.

"Have a go-bag ready to go, with the stuff you need in it. You're not going to have time to pack. And also have a bag in your vehicle, 'cause you may not be at home.

"The last thing is, wherever you're at when this happens, you're going to either be dependent on the people right around you, or you're going to be a resource to those people right around you. So start thinking thataway.

"We can only do so much," he summed up, speaking on behalf of all the first responders: police officers, firefighters, sheriff's deputies. "And quite frankly there's not a lot we can do. At a minimum, get yourself and your family ready.

"That's all I got."

Now one more speaker stepped to the stage: Chief Janik. He was a relative newcomer to town, having taken on leadership of the Gold Beach Police Department less than two years earlier. Immediately his cell phone rang—"Just a sec," he said to an audience that had already sat through almost two hours of presentations and was getting a little restless. He pulled the phone from his pocket.

"Sorry," he said, "it's a call from my wife ... Yes, honey? ... No, I don't know where the hair dryer is ..."

Now people were laughing, getting that it was a gag. "Well, actually I'm at a presentation for a tsunami outreach program ... you bet, and don't call it a *tu*-sunami. It's a *su*-nami ... All right, honey, love you too ... OK, dear, I'll make it brief." And he slipped the phone back into his pocket, to laughter and applause.

He did make it brief. "Almost twenty-four months ago, I really could not pronounce that word," he continued in a measured, authoritative voice, to some sympathetic chuckling. "Having been a desert dweller all my life, it was a foreign concept. However, I continue to learn the enormous impact of this natural disaster. One thing is certainly true. It's not a matter of *if*. It's a matter of *when*."

He encouraged the audience, as the other speakers had done, to form a plan for how they and their families will respond when that day comes, and to prepare to be self-sufficient "not for several hours, not for several days, but possibly for weeks.

"Although it is very hard to grasp the enormity of the problems this will cause us, it would be far more devastating not to have a plan in place to address it," he summarized. "That is something we all can do.

"Now help yourself to some food."

The audience rose and began to break up, and Witter was quickly surrounded by a small crowd. "What do you do if you're in a boat, on the river?" one man asked. Could a person outrace the tsunami if he sped upstream? Witter smiled, adopted a quizzical expression; that was a new one. Racing upriver might work, he said, especially in a jetboat of the type commonly used on the Rogue River, but his best advice was still to land the boat and head to high ground. Another man regaled Witter with his memories of the 1964 tsunami, which he had watched come up the Rogue from his house high on the hill.

A slight woman with short white hair waited patiently for her turn to speak, her husband silent beside her. One stem of her reading glasses was tucked in the knot of a turquoise-and-aqua silk scarf settled just above her parka zipper. "I just wanted to say thanks," she began. She and her husband live up the Rogue River, where they're out of earshot of the tsunami warning sirens, and that had been worrying her. More than once, tsunami alerts from earthquakes around the Pacific had been issued, and she hadn't even known about them. Those alerts hadn't amounted to anything. But what about the Big One, she had been wondering; what if she didn't know about *that* one?

"Now I get it," she said. "I don't need to worry about a siren. I don't need to worry about a warning. Thank you."

34

3/11

Tokyo, Japan, 2011

Young sparrows
Get out of the way! Get out of the way!
A great horse is coming!

—Kobayashi Issa

Kenji Satake was in his office at the University of Tokyo when he felt the jolt of an earthquake's P-wave rattle up through his chair the afternoon of Friday, March 11, 2011. It was the fourth quake in less than a week. Nothing very dramatic followed, even when the S-waves began rumbling through the floor: no books cascading from shelves, no violent wrenching. His building was the newest of the Earthquake Research Institute's three buildings. It was engineered as a base isolation building, with a flexible connection between foundation and structure specifically designed to weather earthquakes. But this quake's shaking was unusually prolonged. Satake knew the institute's director was at a meeting at the Ministry of Education downtown, and Satake's co-deputy director was out of town. Which meant that Satake's immediate concern wasn't the well-being of the 128 million Japanese people. It was the health and safety of the faculty, staff, and students of the Earthquake Research Institute.

One year after pinpointing the date, time, and magnitude of the last great Cascadia earthquake, Kenji Satake had moved back to Japan. Partly it was disenchantment with Michigan—not the university per se, but its setting in the American Midwest. The job required him to teach undergraduate courses in natural hazards and one they called "shake and bake": introduction to earthquake and volcano science. But Satake found it hard to engage the school's Midwestern undergrads. They would perk up at talk of tornados and blizzards but not earthquakes and volcanic eruptions, things that happened to people far away in foreign places like Japan or California. The students had taken a passing interest in the magnitude 6.8 Kobe earthquake that struck Japan's southern Honshu in January 1995, killing more than six thousand people and damaging or destroying some one hundred and fifty thousand buildings, many of them traditional wood frame, tile-roofed homes that had collapsed like stacks of pancakes.

To Satake, the Kobe quake had been of much more than academic interest. *More than six thousand deaths.* Sitting in his office on the other side of the world, Satake couldn't help but think that his skills and knowledge might be better used in service to his own vulnerable country. The Marshall meeting the previous fall had not only resulted in a serendipitous coup, tying the mysterious Genroku tsunami to the last great Cascadia quake, but it had piqued Satake's interest in paleoseismology. With such a long and detailed historical record to rely on, Japanese researchers had only recently turned to analyzing the layers of sediment found on their own country's coastal plain, layers that included sand deposited by tsunamis on the Sendai coast in 1933 and 1896 and 1611, ancient tsunamis such as one in 869 and even earlier, predating written records. He had accepted a research position with the Geological Survey of Japan. Thirteen years later, in 2008, he joined the faculty of the Earthquake Research Institute. Among his responsibilities was participation in a tsunami warning advisory committee for the Japanese Meteorological Agency. That very morning, March 11, Satake had met with the committee to discuss how to improve the nation's warning system.

Satake waited several minutes for his building to stop moving: the duration of the shaking alone spoke volumes to the seismologist. Then he headed to a bank of computer monitors arrayed down the hall from his office. Even before the shaking had stopped, the Japan Meteorological Agency had issued an estimate of the earthquake's size: magnitude 7.9. That was fast, Satake reflected, but was it accurate? And JMA was already projecting estimated tsunami heights: twenty feet at Sendai and the rest of the Miyagi coastline, and ten feet along Iwate, the next prefecture to the north.

Satake began calling around, assessing the damage to the institute. No injuries, apparently, and no obvious damage to the institute's other two buildings, only one of which had been seismically retrofitted. Most people had already evacuated to the courtyard between the buildings, where they now huddled in the freezing afternoon. Satake contacted an architecture professor and specialist in seismic engineering, who agreed to check the ERI's two older buildings for damage. But that would take nearly an hour; until then, those buildings would be off-limits. Traffic outside the campus was gridlocked, and the subway had shut down; it was becoming clear that many people at the ERI would not be going home that night. Satake's building had an emergency shelter room with food and cots, which he now opened as a temporary refuge.

Only later, an hour or more after the first tsunami wave struck the coast, was Satake able to stop long enough to check his computer for updates and to watch the startling video footage of the arriving tsunami. After offshore tsunami pressure gauges had detected a huge tsunami wave moving toward shore, the JMA had doubled its estimate of projected wave heights. But by then, the tsunami had been just one minute out from its collision with the Sanriku shoreline, where many people had already left their homes and were huddled at what they believed, mistakenly, to be a safe elevation.

The JMA had by then also updated its estimate of the quake's magnitude to 8.9, a number Satake found staggering, far higher even than what he and other paleoseismologists in Japan had projected was possible. Only later did he learn the reason for the delay in reporting the much-higher magnitude estimate: the earthquake had been so powerful, it had overwhelmed the country's seismological instrumentation. Scientists in Japan had had to wait until the P- and S-waves, traveling through the earth, were detected by monitoring stations on the other side of the planet. Not for two days would scientists settle on a final figure of magnitude 9, making the Great Tohoku earthquake the fourth largest anywhere in the world since 1900 and the largest in Japan in the one hundred thirty years that modern instruments had been recording quakes there.

Meanwhile, Satake began gathering data to take to a 9 p.m. emergency meeting of the Japanese government's Earthquake Research Committee at the offices of the Ministry of Education downtown. That committee's first responsibility was to provide monthly evaluations of seismic activity in Japan and big-picture projections of the seismic potential of various regions of Japan, but its members always assembled after major earthquakes to provide analysis to government officials. Satake was having trouble gathering complete data from a seismic infrastructure that, despite its sophistication, had apparently been overwhelmed. But the more immediate problem was how to get to the meeting. Mass transit was out, and driving would be impossible. So shortly before 8 p.m., Kenji Satake hunted down a bicycle and set out in the dark, alternately pedaling and walking, dodging stalled cars and passing throngs of dark-suited business people trudging out of the city center on foot.

Among the many topics members of the Earthquake Research Committee touched upon in their meeting that night were reports that the cooling system at one, at least, of two shoreline nuclear power plants at Fukushima had broken down in the wake of the tsunami; officials were apparently preparing to issue an evacuation order for people living in the vicinity. But heading off a nuclear crisis was beyond the scope of the ERC, which that evening was finding itself hamstrung by a lack of data. Power was out over vast swaths of the country to the north, and the tsunami had apparently destroyed some of the monitoring stations linked by underwater cable to the pressure gauges designed to detect and measure it. Regardless, the government was waiting, eager for the cream of Japan's earth scientists to explain how a natural disaster this large could have taken such a well-prepared country by surprise.

Working with what data they could lay hands on, the committee managed to hammer out a statement by about 11 p.m., and Satake bicycled back to the ERI, arriving in time to deliver a status report to the institute's director, back on duty, at midnight. It was about 1 a.m. when Satake got a phone call from a reporter with *Asahi Shimbun*, one of Japan's two largest newspapers. The reporter had requisitioned the newspaper's own plane for a flyover of the Tohoku coastline at first light; he was hoping Satake would accompany him and provide expert commentary. Of course he would go. The tsunami had still been surging at sunset that evening; this flight would provide a first look at what it had wrought, now that the waves had done their damage and withdrawn. Shortly before 2 a.m., the reporter showed up at Satake's office, and the two of them climbed into a taxi headed to Haneda Airport. What normally took an hour by public transit took three hours of crawling on traffic-clogged surface streets.

The plane took off shortly before sunrise, heading northeast, into the sun, toward Fukushima prefecture at the southern end of the coastal disaster zone. The day promised to be as cold and as clear as the day before. The rising sun illuminated the emerald green of the mountain ridges above the coast and glinted off the water below. Water, everywhere. Water where dry land had stood not twenty-four hours earlier. Land that Satake knew well, land he had buried gouge corers in not many years earlier. Just two years before, he had helped produce the Earthquake Research Committee's long-term forecast, suggesting an 80 to 90 percent probability in the next thirty years of a magnitude 7.7 earthquake in the central Sanriku area, the location of the epicenter of yesterday's quake. Clearly, that projection—ominous as it was—had actually significantly underestimated the risk.

The shoreline had subsided, just as the research had indicated it would. Land that had been dense with houses and buildings of all kinds—businesses and schools and hospitals—was now dotted with the steel bones of reinforced buildings stripped of their walls and surrounded by acres of empty concrete foundations. Buildings stood askew, next to ships resting on their keels. Outside of the cities Satake saw bending ribbons of brown leading into the mountains: river valleys that should have been a patchwork of green and yellow farmers' fields.

The sky was crystalline, granting a nearly unblemished view but for the clouds of smoke puffing up from still-burning fires. A quiet, cold morning, smoke curling upward, no cars moving on the broken, sand-covered roads below, and a wide swath of driftwood and other debris bobbing on the sea's surface and stretching nearly to the horizon. It could almost be the year 1896, or 869 for that matter, Satake reflected, but for the oil slick covering the surface of the sea.

There would be plenty of finger pointing, and there should be, given what scientists had learned and the government had disregarded from the paleoseismology studies, Satake figured. Hazard mitigation steps could have been taken, lives could have been saved with more education, more stringent building codes, a firmer hand with the nuclear power industry. But ultimately it wasn't humanity's fight Satake was witnessing out the airplane window. It was the result of a long-running feud between tectonic plates. *We are merely collateral damage in an entrenched battle that has spanned generations, empires, millennia, far older than the existence of humanity itself.* It looked quiet now, from 2,000 feet up, but Satake knew the tension was still there. The plates were still moving. The land was changing every moment.

It was half past midnight on March 11 in Seaside, Oregon, when Tom Horning saw the revised magnitude figure for the Japanese earthquake appear on his computer screen: 8.9. Now it was all starting to make sense, the quake's magnitude matching the severity of the tsunami that Kirsten and Chamber of Commerce director Al Smiles had been watching on TV, that he himself had seen on CNN. He leaned back in the oak office chair, stretching his back and giving his eyes a moment's rest from the screen. Magnitude 8.9: nearly as big as the last Big One to strike the Pacific Northwest coast three centuries earlier: 311 years, one month, and two weeks ago, to be exact. *The one no one knew anything about until twenty-five years ago,* Tom mused. *Well, the Indians did, but nobody listened to them.*

Now what had been an odd sort of evening—a curious alert, a few phone calls interrupting and abetting Tom's procrastination—turned into a frenzy, as if that number itself—8.9—had magical properties. A reporter from a Portland television station called: the station was sending a crew to Seaside; could Tom do a live stand-up interview at dawn? Again the computer chirped: the alert status for the coast of Oregon and most of California had jumped up to *warning*—the highest alert level in NOAA's vocabulary, urging immediate evacuation from low-lying coastal areas. Al Smiles, on the phone again, still watching live and instant-replay footage from Japan: "You've got to see this, Tom! Oh my God ..." Kirsten, checking in: "We can't leave Dad here; what do you think, Tom?" Kevin Mitchell, Tom's neighbor at the end of Neawanna Street: "Kirsten said you were at your office; time to get the hell out of Dodge, eh Tom?"

"Maybe, maybe, Kevin," Tom responded. "I don't think it's going to be that big."

Through the wall, Tom could hear a NOAA weather radio squawking. He walked across the hall, unlocked the door to Tsunami Deb's office, and stepped in to listen: *tsunami coming, evacuate to high ground.* The phone was ringing when he returned to his own office: a second Portland television reporter seeking an interview. *These distant tsunami alerts: They get people thinking about tsunamis, and that's good,* Tom mused. *But it gets them thinking about them the wrong way.* People mix up local and distant tsunamis, confusing a genuine emergency requiring instant action with a mere call for prudence, a suggestion to move away from the water's edge within the next several hours. The confusion was understandable; he knew that. No one living had ever felt the terrible power of a magnitude 9 earthquake on the Oregon coast, nor witnessed the tsunami that followed.

Sometime after 2 a.m. the phone calls started to ease off, giving Tom time to focus on the new task he had set for himself: making his own projection about the size and shape of this distant tsunami, taking into account the weather and tides, history and prehistory, data still arriving from Japan, the distance from the source quake, and the angle of the fault, all of which suggested more mischief in southern Oregon and northern California than here in northern Oregon. Both his home and office were perched no more than six or seven feet above mean high tide, putting them in a "low-lying coastal area" and worse, directly east of the bay mouth, where even small tsunamis can get ramped up. But Tom had faith in his own calculations. It looked like the first tsunami wave—it was never just one wave but a series, and the first wasn't always the worst—might reach the northern Oregon coast as early as 7:20 a.m. That would put it here on an ebbing tide, about halfway to a 1.3-foot low tide. The surf was high—that would add height to an incoming tsunami—but it had been dropping through the night. All things considered, Tom figured the tsunami from Japan wouldn't be all that high: two or three feet, six at most, nothing to get too worked up about, if history was any guide—and it was. The 1964 tsunami from Alaska had been some eighteen feet high when it rolled in on a tide that was at least four feet higher than it would be at 7:30 the next morning. This tsunami wouldn't be a dud, Tom figured, but neither was it going to be even a minor disaster, not in Seaside. A very full, churning Necanicum River is what he expected— probably nothing more than that. Payback for the tsunami Oregon sent to Japan three hundred-plus years ago? He smiled at the thought. *Not quite.*

None of Tom's siblings had called—which he appreciated; those who cared enough to call knew he'd be busy, knew that if anyone had a plan at a time like this, he would. Which he did. First, finish up those meeting minutes. Next, take a nap on his office couch. Then get up and meet that television reporter at the high school.

Kirsten called again, nervous; the warnings on TV about a tsunami hitting the Oregon coast from Japan were pretty doom and gloom, and she wanted to hear Tom's own assessment of the danger. "I don't think you need to worry," he said, ticking off his calculations. "It's not going to be any higher than a typical winter high tide."

Tom never did get that nap. Right around 6 a.m. he finally wrapped up the land trust minutes and locked his office door behind him. It was still dark as he drove the scant mile to the overlook west of the high school to meet the television crew. Chilly, but at least it wasn't raining. Tom, in blue jeans and waterproof parka, was his usual garrulous self; he knew his material and he loved to expound. He'd spent most of the night studying this event's particulars, so he was easy with the TV crew, standing there in the dark, a light wind from the south ruffling his thick mop of graying brown hair. "It's probably not going to be very big," he told them. "The only place it'll be a problem is along river mouths and inside of bays. Even on the beach it shouldn't be too much of a problem, provided that it's not a rocky, steep beach." Just be cautious, he urged anyone listening. "Don't take any unnecessary risks. Treat it like a high tide." The whole interview took less than five minutes.

At home, Kirsten was in the kitchen, spooning leftovers into plastic containers: provisions, in case she got stranded (or, more likely, for lunch at work later that day). Tom settled in front of the television, which was broadcasting tsunami news nonstop, and began watching, uninterrupted, the images coming out of Japan.

He was stunned. Riveted. It was one thing to hear Al Smiles's breathless play-by-play on the phone, one thing to view the still photos of a tsunami's aftermath that he'd been parading in front of Seaside's city council for the past decade or more. It was another to see, in real time, footage of a kind and number and variety no one had ever captured of a tsunami: the relentless flow, the crushing destruction. He and the rest of the world had seen the shaky images tourists caught of the Sumatra tsunami back in 2004, but here was a tsunami in a developed country where it seemed everyone had a smartphone capable of video, not to mention journalists in helicopters capturing the wave even before it hit land, when it was still a long, white line marching toward the shoreline, chillingly relentless. It was astonishing and horrifying: houses exploding, houses floating, houses turned into battering rams smashing other houses. Fishing boats sailing down what had been city streets, boats and minivans pouring over seawalls that had become nothing more than gray waterfalls. Black, debris-laden tongues of water encroaching unceasingly upon a neat patchwork of farmers' fields, the water piling up and churning restlessly at elevated roadways. Survivors clustered on hillsides, pointing and sobbing, the lucky ones. Horrifying for those caught in the maelstrom there. And edifying—or should be, Tom hoped—for anyone paying attention in the Pacific Northwest. Cities built upon a flat coastal plane, like Seaside and Gearhart and Warrenton. A fault zone right offshore, where two tectonic plates collided. Just like Seaside.

Holy mackerel, Tom thought, then "Holy mackerel," he said out loud. "Holy mackerel."

And then, in a rush, that morning's dream came back to him, the one that had awakened him with a start at 5:30 a.m.—actually, he realized, it was the previous morning, twenty-four hours ago, the morning before this past sleepless night. It wasn't Tom's first tsunami dream; he seemed to have one, or remembered having one, every few years. Usually they were simple remixes, slightly exaggerated in scale or hue, of his own childhood tsunami memories: a sand-covered lawn strewn with gigantic logs and seaweed, the tangy smell of low tide, the cold, clammy feel of the air at midnight. Comforting dreams. This was different: a vision not of a magical night in the past but of an apocalyptic future.

In the dream he was standing in his backyard a few steps from his house, looking across the estuary to Little Beach at Gearhart, when he had an inkling that a tsunami was coming. A certainty, in fact: the Big One. His dream-self shifted his gaze south, toward the sand dune north of the bay mouth, and behind it he could see the ocean rising up behind the dune about a mile offshore. The growing wave front bulged and was covered with faint three-dimensional grid lines, a grid that was expanding as the wave grew, swelling like an inflating balloon. And the wave wasn't ocean-colored, wasn't gray or blue-gray or gray-green, but pink—the ruby pink of a pink grapefruit. He watched the pink wave grow, bulging like a bubblegum bubble, like the bulbous throat patch of a magnificent frigate bird. As he watched, the fat wave just kept rising—fifty feet tall, one hundred, two hundred, finally three hundred feet—the dream was very specific. A wave front hovering three hundred feet above the beach, gathering itself into a furious pink balloon of destruction until it was towering over all the oceanfront houses in Gearhart, blotting out the sky to the west.

In the dream, watching the wave swell, Tom knew he needed to leave, to run east as fast as he could go, east over the bridge across the creek to the forested hills, east to elevation and to safety—wasn't that what he kept reminding his fellow citizens to be prepared to do when the Big One hits? To not wait to gather supplies or clothes or treasures or even loved ones, certainly to not stand and watch, but to go to high ground, as high and as fast as you can? Instead, he just stood there, frozen in place and staring, his eyes bugged out, his jaw slack with terror and awe. And as the wave finished gathering itself and was about to collapse forward, Tom, frozen in place in his yard at the mouth of the river, took one long, deep breath—

And that's when his eyes had shot open and he was awake, hyperventilating and staring at his bedroom ceiling in the dark before dawn. *Holy mackerel.*

On the TV, the picture had switched to bluish footage of a beach in Hawaii at night—waves lapping regularly and splashing calmly on the gradual rise of sand, documentation of the tsunami's minimal impact there on its trip across the Pacific. Clearly the tsunami had not spread mayhem when it hit the mid-ocean island chain on its journey east and south. And if it wasn't slamming Hawaii, it surely wasn't going to have a big impact on the West Coast of North America. Then the picture switched to Tom's own smiling face, hands in and out of his pockets as he explained to the reporter the dynamics of this tsunami and why he expected its impact on Oregon's north coast to be minimal. He was immediately followed by news anchors urging their viewers to stay tuned and stay panicked, or so it seemed: to not be misled by the jovial geologist who clearly was underestimating the dangers ahead. He chuckled, watching. *So predictable,* he thought. *Every tsunami alert gets the media so riled up. It's like they still don't understand. The Big One is coming. But this isn't it.*

Kirsten, like Tom, had stayed up all night, and now they began talking again about next steps. Her plan was to leave the house by about 6:45 a.m. and head up to the hillside neighborhood at Thompson Falls, across Neawanna Creek: out of harm's way, but with a view, in case anything interesting should happen. Plenty of people would be driving up Crown Camp Road toward the old logging company headquarters—they always did in a tsunami alert—and she thought she'd avoid that traffic jam. After the excitement had passed, she planned to continue north on U.S. 101 to her office at the college in Astoria.

Tom reiterated his own plan. He was confident the tsunami wouldn't be high enough to touch their house, but with evacuation sirens blowing, they couldn't in good conscience leave Mike alone at home. So Tom would take Mike with him, first to the bakery in Gearhart for coffee and pastries, then to the second-floor deck off a friend's house across the street from the high school: prime tsunami-viewing.

Tom's father-in-law was his usual cheerful self as Tom helped him on with his coat and pointed him out the door, toward breakfast and the tsunami; Mike didn't quite understand what was going on, but he always enjoyed getting out. Tom glanced at his watch: plenty of time. It really shouldn't be much of anything. But every tsunami is different, with its own signature, its own personality, its own lessons to teach. You never knew exactly what would happen until it happened.

He could hardly wait.

As soon as he felt the aftershock's P-wave arrive, like a jackhammer under his chair, Chris Goldfinger did as the other patrons in the airport restaurant were doing: he laid his paper napkin over his *katsudon* to keep plaster dust filtering out of the ceiling from dropping onto his dinner. Aftershocks such as this one had been almost continuous in the two days since the big quake: not so much discrete events as a way of life, the new normal. He waited as the quieter, rolling S-waves and surface waves passed through the floor and his chair and his own body. Then, like the other customers, he removed the napkin and returned to slurping his pork cutlet rice bowl. Most of the restaurants in the international terminal of Tokyo's Haneda Airport were out of food; Goldfinger had been happy to find one still open and was content to accept the sole menu item it offered. He still had two hours before his midnight departure; who knew if there would be any food available on the ten-hour flight back to the U.S.?

Nothing more useless than an earthquake scientist right after a huge earthquake, Goldfinger mused—a foreign marine geologist, at any rate. In the weeks ahead, scientists from around the world would be descending on the Tokohu region's coastal plain, joining Japanese scientists documenting the tsunami's run-up. Right now, Goldfinger knew, the best thing he could do for the Japanese people was to leave the country. They had enough to do locating and sheltering and feeding survivors and dealing with what was shaping up as a nuclear catastrophe; the last thing they needed was another foreigner to feed and lodge.

Two and a half days earlier, Goldfinger had been at a scientific meeting at the University of Tokyo's Kashiwa campus, northeast of the city center, sitting through a day's worth of research findings on the 2004 Sumatra earthquake and making a presentation of his own. When Goldfinger felt the jolt of a P-wave strike the building, he was less apprehensive than annoyed; they were already running behind and couldn't afford another distraction. Certainly no one in this gathering of world-class earth scientists was alarmed, not at first. In the five days since Goldfinger had arrived in Japan, there had already been two sizeable earthquakes, one of them magnitude 7.2, triggering tsunami warnings; it was starting to feel routine. The speaker had paused, waiting for the rolling of the S-waves to arrive and to stop. But a minute passed, and the building was still moving. Even then, even before the shaking stopped and he understood the severity of this event, Goldfinger sensed he was experiencing something exceptional; he could almost feel the plates grinding together under his feet.

"Let's go outside!" someone said, and everyone did, gathering on the plaza. Another minute passed; Goldfinger watched a flagpole atop a seven-story building whip back and forth and watched the base isolation building he had just exited roll to and fro atop its foundation. He heard no roar, just the rattle of dry leaves in a row of trees at his back. People were checking their watches—four minutes, five minutes. Finally, with the Earth still shaking, Goldfinger proposed that everyone pose for a group photo. In that deceptively still image, nearly everyone was caught smiling broadly. It's not often that an earthquake of this size strikes in the middle of an earthquake scientists' meeting. No theory of earth science had ever projected that the plates upon which Japan sits were capable of producing a quake the size of the one that this group was witnessing.

In the years since that first cruise on the *Melville* in 1999, Goldfinger had gained near-star status in geology circles for his turbidite records. He'd followed that cruise with another in 2002, gathering more cores at Rogue Canyon and Hydrate Basin off Oregon to bolster the story. Those cruises were followed by more fieldwork, in Cascadia and beyond. In Phuket, Thailand, at a gathering on the eve of a research cruise in the Indian Ocean, Goldfinger had been challenged by a crew member to predict the pattern he would find in the offshore turbidites responding to the 2004 Sumatra quake. No problem, he said, quickly sketching the pattern on a whiteboard. It was precisely the pattern they did find in the cores pulled days later from the ocean floor. His turbidite methodology was so reliable, he could use it as a parlor trick.

Everything he and Nelson had deduced that July day in 1999 aboard the R/V *Melville* had held up upon further inquiry; in fact, the story just got more credible as he worked up the data. With radiocarbon dating of the foraminifera—fossilized plankton—found in the sediment immediately below each turbidite, he had identified a sequence of forty earthquakes in a ten-thousand-year period. Forty ruptures of the Cascadia Subduction Zone spanning nearly the entire Holocene epoch. The record began perhaps five thousand years after the first humans crossed the Bering land bridge, not too many millennia before the first Paleo-Indians settled the Pacific Northwest coast, a coast that only occasionally but devastatingly was wiped clean by giant tsunamis triggered by giant earthquakes (only two of those quakes appeared to have been smaller than about magnitude 8).

But only about half of those forty earthquakes registered in the turbidites appeared to have been complete ruptures of the entire 620-mile-long subduction zone. Twenty-one of them were partial ruptures of just the southern end of the zone: south of Astoria, or Newport, or Gold Beach. That meant that the sites of present-day Vancouver and Seattle, for instance, had been seriously shaken only nineteen times in the past ten thousand years, that the Makah or their predecessors at Washington's Neah Bay would have seen the sea flood over Waatch Prairie only once every twenty generations, on average. Meanwhile, the Yurok, Tolowa, and Wiyot people living at the mouths of the Eel and Klamath and Smith and Chetco rivers far to the south would have felt the Earth convulse and watched prairie become ocean twice as often. Potentially twice in a single lifetime, according to the turbidite record, which suggested one recurrence interval as short as twenty-seven years, though the average interval along the southernmost segment of the rupture zone was closer to two hundred and forty years.

It's so obvious, a five-year-old could do it, Goldfinger still liked to say, though proving to the scientific community's satisfaction that the turbidites really did spell out the seismic record wasn't quite that simple. There was high-tech radiocarbon dating adjusted for every possible quirk or bias, shipboard imaging to capture the colors of the sediment layers immediately after splitting open the cores, even x-ray and CT scanning to reveal the subtle details in soft muds. But there was no substitute for going into the cold core locker at the edge of the Oregon State University campus and pulling out a core, inhaling that musty smell of ancient mud and eyeing the sediments themselves.

One of the points of nailing down frequency and size of previous Cascadia quakes was to make an educated guess about how soon the next one might occur and how big it might be, and that's where it got sticky. For one thing, it appeared that the Cascadia quakes came in clusters. Not clusters of a few earthquakes in a week or a month, but two to five quakes bunched a few hundred years apart, then a gap of eight hundred to a thousand years before another. And the last quake in such a cluster tended to be a biggie.

Was the last quake in A.D. 1700 the fourth in a cluster of five, or the last one we'll see for a millennium? At magnitude 9, that quake was big but not the biggest; seven of the quakes in the record were judged to be even larger. Maybe that was it for a while, which would mean it might be—according to the turbidite record—four hundred and forty to eleven hundred years before the next one: generations from now, maybe dozens of generations. On the other hand, maybe this cluster—if it was one—might kick out one more before a long recharge. And those figures were just for a full rupture. Goldfinger had little doubt, from looking at the data, that a partial rupture on just the southern end of the fault was due. Overdue, in fact.

What are the chances? That's what it always came down to for the public. With no reliable way to predict exactly when an earthquake will occur, people at least wanted to know the likelihood of it happening in the next few years, or in their lifetime, or their kids' lifetime. There were a lot of ways to look at it, but using what was widely agreed to be the preferred methodology, Goldfinger figured there was a 7 to 12 percent chance of a full rupture, all six hundred and twenty miles, in the next half-century, with odds on 12 percent. As recurrence intervals shortened toward the southern end of the subduction zone, the likelihood of a quake there in the next fifty years increased. Goldfinger estimated an 11 to 17 percent chance of a rupture just from Astoria Canyon (near the mouth of the Columbia River) south to the mouth of the Klamath River or all the way to Eureka. The odds of a rupture limited to the area from about Newport, Oregon, to northern California rose slightly to 15 to 21 percent. And the chance of just the southern 137 miles of the zone rupturing—roughly, everything south of about Gold Beach—was 37 to 43 percent.

In Japan, as he worked at arranging a new flight home, Goldfinger found himself reflecting on the parallels between this Tohoku quake and the last, and next, Cascadia event: the lessons North Americans could take away and he could share from his experience of being a geologist two hundred and thirty miles from the epicenter of one of the biggest earthquakes in history. He'd already shared some thoughts with American reporters who had tracked him down in Tokyo and interviewed him in his hotel room via Skype.

"It seems that the more 'advanced' a society becomes, the shorter its memory," he would later write in a blog. "The Andaman Islanders did better in the 2004 quake than anyone else," he asserted, referring to a remote tribe of hunter-gatherers in the Indian Ocean who, trusting the dictates of tribal oral history thousands of years old, reportedly moved inland to higher ground at the first rumblings of the earthquake. "Native Americans not only have a memory of the last Cascadia earthquake 311 years ago, they have a memory of the explosion of Mount Mazama about 7,600 years ago. We, on the other hand, can't remember much that happened before Twitter and Facebook. Modern societies intentionally discard anything old. So instead, we have to rely on science, not social memory, to fill in what we have forgotten."

Since the earthquake, he would write, "I've had the strong feeling that we're making many of the same mistakes the Japanese did. While they were the best prepared in the world, they have the same political and economic pressures and make the same compromises we do. I think if we want things to go any better in the Pacific Northwest, we have to take a bit of a different tack and speak more clearly about what the problems are and how to address them."

Rescheduling his flight home had been a scramble, but here at Haneda, actual airport operations seems to be well in hand; everything pointed to an on-time departure. Goldfinger, sitting in his window seat, could feel the plane bouncing a little while baggage was being loaded into the belly of the aircraft. Then he heard the baggage door slam closed. But the bouncing sensation continued, and out his window he could see the jet's wing tips fluttering in the dark, like the quick wing beats of a duck. Then there was a brief lull before the plane started taxiing down a bumpy runway that, just days earlier, would have been smooth as silk, freshly paved as part of a recent airport upgrade. Not until the jet was airborne, detached from the tics and spasms of the restless Earth's crust, did Goldfinger's body begin, for the first time in days, to finally relax.

35

Strategic Investment

TOM, 2012

THE SKY ABOVE THE DARK MASS of Tillamook Head looked bruised, the slate gray broken here and there with slivers of pale blue and shards of sunlight, as Tom steered the Accord into the ball fields parking lot off Wahanna Road. The stubborn leaves of the blackberry thicket edging the parking lot were only beginning to turn yellow, though the alder trees beyond the brambles were mostly bare. Tucked in the southernmost corner of the parking lot was a large drilling rig on caterpillar tracks, roaring and vibrating and clanging as it pounded a hydraulic ram through a hole in the pavement. Tom, in faded blue jeans and a red hooded Seaside High Football sweatshirt, strode toward the figures gathered around the rig, greeting Sean Dixon, the geologist in charge: a younger, slighter man in yellow waterproof overalls and a gray wool sweater. Conversation, however, was impossible against the din. A drill operator reached toward a switch to power down the rig.

"If you could get to twenty-five feet, I'd be happy," Tom said.

"What depth do you want to start sampling?" asked Dixon.

"At five feet or so," Tom replied, "once you know you're out of fill. When you start getting into black organic stuff, you'll be in the marsh. I'd like to start at that point."

"We'll keep an eye on the cuttings."

"Good!" Tom straightened up and smiled. "I don't know enough about the details of finessing these kinds of things, so whatever you think is best. I really appreciate this."

"I'm interested too," Dixon said, signaling to the driller to resume pounding. Dixon worked for Geocon, a leading West Coast geotechnical consulting firm hired by the Seaside School District to confirm that the hillside site Superintendent Doug Dougherty was eyeing for his proposed consolidated school-district campus was rock-solid and not prone to landslide in a big earthquake. Dougherty had asked Tom to consult on the project as well. Tom had already spent a day guiding Dixon through the tangle of logging roads in the hills above town, observing as Dixon installed inclinometers at the proposed campus site to detect any slope movement over the winter, kibitzing as they probed the bedrock underneath the ball fields parking lot at the base of the hill to assess how well it buttressed the slope. Tom suspected Dougherty had hired him in part out of charity, throwing a little paid work Tom's way to thank him for hours of free consulting in service of what was shaping up as a monumental and unprecedented project: swapping five schools and their associated campuses in three towns for a single, brand-new, seismically engineered, tsunami-safe school campus. But Tom knew that Dougherty wasn't just throwing him a bone. No single individual knew more about the geology of Clatsop County and the history of paleoseismology research there than Tom Horning, Seaside High School Class of '72.

The driller, backing the auger out of the hole in the pavement, detached a two-foot-long metal cylinder at the end and handed it to Dixon, who gave it a shake, then unscrewed the top and separated the two halves of the split spoon sampler, laying them face-up on a metal tripod bench and gesturing to Tom to have a look. The two geologists were actually looking for different things: Dixon for evidence that this hill will hold in a quake, and Tom for cobbles, silt and sand layers, or other tracks of past tsunamis. Why not, while they were here drilling anyway, piggyback some science onto an engineering project?

Tom—silver-haired, but still a "barrel on stilts," as his sisters used to tease him—bent over the core sample, half-filled with muddy sediment, peering through tarnished wire-rim glasses with the discernment of a vintner poised for his first sip of Beaujolais nouveau. He poked at the muck with an index finger, then he plucked a folding knife from one pocket of the tan utility vest draping his sweatshirt and began smoothing the face of the core sample with the blade, picking at it with the tip. "Teeny microscopic creek sand," he narrated to himself. He flicked out a brown sliver. "A piece of wood, what looks like a mud flat. A molten Hershey bar!" he pronounced, succinctly capturing the core's appearance. With the tip of the blade, he nicked out a sample of sediment and popped it into his mouth, probing it with his tongue to assess the grain size. Then he spat it out onto the asphalt.

THINGS HAD LOOKED SO PROMISING in Seaside a decade earlier. By the end of 2003, four of the city's twelve bridges had been rebuilt, two over Neawanna Creek and two over the Necanicum River, giving people in the center of downtown and those a half-mile north a clear path for evacuation, up Broadway and Twelfth Avenue. The Tohoku quake had dramatically demonstrated how important solid bridges were in a tsunami: thousands more may have died had Japanese bridges not held up as well as they did. Even if they crumpled a bit, if they lost a railing or slumped on one side and were unable to carry car traffic, Tom was confident Seaside's four new bridges would at least allow pedestrian evacuation, which was the kind of evacuation everyone needed to be prepared to do.

Then it had stopped. It had been nine years since the last bridge was rebuilt. That left eight bridges, most of which were more than fifty years old and unlikely to survive the four or five minutes of severe shaking geologists predicted. People in his own neighborhood and, worse yet, at the south end of Seaside would have a hard time getting to high ground fast enough.

A few months earlier, the city had completed a project to fill and stash one hundred and eight fifty-five-gallon drums of emergency supplies in the garages of private houses up in the hills east of town, above the inundation zone. The supplies inside—food, water purification tablets, radios, tarps, and the like—were paid for with funds from the same federal grant that had installed Seaside's voice-capable tsunami warning sirens in 2007: two items checked off the city's tsunami to-do list. But the sirens would only be helpful in the event of a distant tsunami; with power likely to get cut, they might not work after a Cascadia earthquake, and as Tom often reminded his fellow citizens, the earthquake *is* the warning. And what were known locally as the "blue barrels"—repurposed syrup containers donated by two local candy shops—held only enough supplies to feed two thousand people for three days: a drop in the bucket, by Tom's calculation. What the town really needed, he figured, was enough food for twenty thousand people for twenty days. He had a grander vision, of shipping containers filled with more food, more equipment, and emergency shelters parked at Crown Camp: a Tsunami Evacuation Park, if you will. Meanwhile, Tom agreed, the blue barrels were better than nothing.

He was also a little frustrated with his fellow Tsunami Advisory Group members. Certainly they were well meaning, but sometimes it seemed more like a social club than a group with a life-or-death mission. No doubt they were victims of warning fatigue, a malaise the whole town seemed to have contracted. The TAG would hold tsunami preparation fairs or sponsor talks from time to time, and every time the number of participants who showed up seemed to dwindle. Tom figured half or more of the motel clerks in town weren't even aware that the city had tsunami evacuation maps, much less how to put their hands on one. Seaside had lots of low-wage jobs and, as a result, a fairly transient work force. Newcomers wouldn't have a clue what to do in a Cascadia quake. Nor would many of the tourists they catered to. Hotel owners seemed to shy away from tsunami preparedness, worried that talking about it would scare away business. Tom didn't get it. Rather than pretending the risk doesn't exist, why not do something about it and claim bragging rights? "Seaside, the Most Tsunami-Ready City on the Oregon Coast!"

There would be little up-ticks in interest, such as after—immediately after—the Tohoku earthquake in Japan. The very morning Seaside was bracing for that distant tsunami (that wound up creating not a ripple in the Necanicum estuary), the Seaside City Council happened to be holding a goal-setting session to set priorities for the next two years. Suddenly tsunami preparedness, which until then had seemed to be barely within council members' consciousness, made it onto the list of goals for the next biennium. The eighth goal of eight, but on the list nonetheless. But what good is a goal, Tom mused, if you never take concrete steps to achieve it, or if you word it so vaguely that you don't have to?

The city manager was well-meaning; Tom understood he had a lot of competing priorities. Like Tom, Mark Winstanley had grown up in Clatsop County and had his own memories of the 1964 tsunami. Both of them had graduated from Seaside High in 1972 and headed to OSU, where their paths diverged, Winstanley pursuing a business degree. Both had long since returned to their hometown. They'd remained friendly; Tom had been an acolyte at Winstanley's wedding. But every time Tom challenged the city to spend some money on tsunami-safe infrastructure, Winstanley seemed to have a pothole that needed filling, or a flat tire on a fire truck.

Kevin Cupples, the planning director, had spent his entire career on the Oregon coast, working in Coos Bay before he took over planning in Seaside. Every year he applied for state funding to get another bridge rebuilt. Every year he got turned down—the ranking system had changed to one less favorable to Seaside than it had been a decade earlier. You couldn't say he didn't try.

"I don't care what's going on downtown," Tom would rant to his morning cronies at Pacific Way Bakery. "Maybe you take a day off to go quell a riot or something, but every year the question should be, how many hundreds of thousands of dollars of our budget is going to go for tsunami purposes? Even if it doesn't hit for two hundred years, it's the best investment we'll ever make."

Which was the thinking behind the Tsunami Strategic Investment Plan Tom had begun working on in 2008 and had been refining ever since. Nearly five years of work, off and on, distilled into twenty-three pages heavy on graphs and charts and annotated photographs, with text organized in bullet points to facilitate understanding, beginning with a blunt introduction. "Seaside will be largely destroyed by a Cascadia earthquake and ensuing tsunami at some point in the future," the opening summary began. "Disaster may strike sooner than statistics might imply."

At twenty-three pages, the plan was a fraction the size of his 402-page master's thesis but, as Tom saw it, potentially six thousand times more important: once for every resident of Seaside whose life it might help save, not to mention the additional twenty or thirty thousand tourists crowding into town on the busiest days of the summer. The subtitle summarized the plan's recommendations: "Education, Evacuation, and Survival: 20,000 refugees for 20 days by 2028." Explain, to locals and tourists, over and over, what to do and where to go when the Earth starts shaking. Give them the means to get to high ground: that's where the bridges come in. Cache supplies to keep them alive for three weeks while overwhelmed emergency personnel deal with broken highways and bridges and collapsed buildings not only on the coast but inland. Tom knew none of this could be accomplished with the wave of a magic wand. So he was aiming twenty years out—2028. There was a reasonable chance the tsunami will not have hit by then. In the meantime, at least city officials could sleep at night knowing they were making concrete progress.

And Tom made specific recommendations. The middle of town was in fairly good shape, evacuation-wise, thanks to the four rebuilt bridges. But the densely populated south end of town remained particularly vulnerable. Fixing the Avenue U and Avenue G bridges over the Necanicum River, at $3 million or so each, would cut evacuation time for people there from about thirty-six minutes to twenty: a survivable threshold. At least it would be a start.

The problem, Tom saw to his growing frustration, was that four of those twenty years had already passed, almost no progress had been made, and city officials seemed to be sleeping just fine. It was unconscionable, as far as Tom was concerned, for someone who possessed this information to fail to act. "You can't deny the risk of this thing!" he would complain to the city council in public meetings and to Winstanley and Cupples privately. "You've got to act as through it's imminent, because it if turns out it's sooner rather than later, the blood is on your hands!" *You may as well be running a concentration camp where you exterminate life*—that's how Tom saw it. He didn't say so out loud; people might take it the wrong way. But the issue was that stark to Tom. You had the choice to save people, and you chose not to.

Say one human life is worth somewhere around $1 to $4 million, he'd posited to the planning director: that seems to be what juries think. Call it $1 million. It's going to cost $40 million to fix the town: to rebuild the remaining eight bridges and stash adequate survival supplies, plus a little money for on-going education. Spread that $40 million over twenty years, that's $2 million a year. In the worst-case scenario you could lose five to ten thousand lives in Seaside alone; that's $5 to $10 *billion* in lost lives. It makes $40 million look like chicken feed.

Take another random disaster: fire. Like a tsunami, you never see it coming; you know it's going to happen somewhere, sometime, to someone, but you never think it's going to be you. The City of Seaside spends a half million dollars every year supporting the volunteer fire department: a handful of paid staff, plus facilities and all those trucks and equipment. All that to save one life every five years: that, Tom had calculated, was about how often someone dies in a house fire in Seaside. Twenty lives in one hundred years. In the next hundred years, there is a good chance one thousand times that many people will be running for their lives from a tsunami in Seaside. Helping to save twenty thousand lives: that would be getting a bang for your buck. It was a simple matter of social justice.

He understood the problem. Residents of the Gulf Coast know that every year they'll have a damaging hurricane and every hundred years or so a devastating one. Floods, tornados, even earthquakes in quake-prone places such as California: people remember these, and they let those memories inform their decision making. But when the last tsunami was more than three hundred years earlier, it required people to engage in a highly creative act: to create memories of a future that has not yet come to pass, and to believe in them, and to act on those beliefs.

Case in point: Hurricane Sandy, the superstorm that had devastated parts of New York City, New Jersey, and surrounding states a month earlier. Follow-up news reports revealed that politicians in New York had been warned, decades earlier, to take steps to prepare for a storm many times the usual size. "I don't know that anyone believed," a still-reeling Governor Andrew Cuomo told news reporters a month later, as recovery efforts dragged on. "We had never seen a storm like this. So it is very hard to anticipate something that you have never experienced." The death toll from that storm, a devastating event that had riveted the world's attention: about 113 people, spread over several states.

One recommendation from Tom's Strategic Investment Plan was finally on the table; earthquake-proof pedestrian bridges. So far, so good: tsunami evacuation will have to happen on foot, and you can probably build two pedestrian bridges for the price of one road bridge. What puzzled him were the criteria the city was using to determine where to site those bridges, criteria such as facilitation of foot traffic between neighborhoods and commercial areas, access to recreation and scenic natural areas, and tsunami evacuation. To Tom, that last criterion was the only one that really mattered. Why would you spend any money on a pedestrian bridge in Seaside, Oregon, if it didn't help more people get to high ground faster? Tom liked hiking and bird watching as much as anyone, but he couldn't understand why public safety wasn't job number one. As Cupples explained it, if Seaside wants the state to go halves on a new bridge, the city has to play by the rules and site its pedestrian bridges according to priorities set by state transportation officials. But wasn't that putting the cart before the horse? Tsunami preparedness in a town like Seaside, it seemed to Tom, called for extraordinary measures and visionary leadership to adequately prepare for what could be—undoubtedly will be—the worst natural disaster in terms of lives lost in the history of the United States. What Seaside had were two earnest bureaucrats, doing their best.

The school district was the one bright spot. Superintendent Doug Dougherty understood the risk, and he wasn't waiting for someone else to deal with the problem or for funding to magically appear. He was moving full speed ahead with his ambitious plan to relocate the entire school district on forestland east of the city limits. Weyerhaeuser was asking $40,000 an acre for the forty-nine-acre site Dougherty had identified; school district lawyers had drawn up an offer for $1,800 an acre: the assessed value. Dougherty was prepared, with the school board's blessing, to start condemnation proceedings if necessary. The plan for the campus had changed, based on engineers' recommendations. Rather than separate buildings for elementary, middle, and high schools, engineering consultants were urging the district to build a single "immediate occupancy" building—meaning, a building so sturdy, so well engineered for seismic safety, it could be occupied immediately after a magnitude 9-plus earthquake without so much as an inspection. Such a building on that hillside would require a beefy foundation with piers perhaps five feet in diameter buried deep in the earth. Multiple buildings engineered to this standard would be prohibitively expensive—hence the revised concept of one multi-story building for kindergarten through high school, or at least a combination gym and cafeteria that would double as a community emergency shelter. One week earlier the district had hosted representatives of seventeen architecture firms from around the country who were interested in Seaside's plans for a tsunami-safe campus. Dougherty planned to award a contract to one of them by the first of the year and to ask voters in November 2013 to approve a bond measure to pay for it.

Cannon Beach Elementary School was a little more problematic. Two engineering consultants had independently assessed the structure and found that its Quonset hut-style gym was likely to collapse in a big quake, as would an old brick chimney and the classrooms' unreinforced south wall. A proposal was made to close the school, but parents protested, organizing a Save Our School group. Word from the state geologist that there was a roughly 15 percent chance of a tsunami hitting town in the next fifty years seemed to reassure rather than alarm some parents. "What if I told them I planted a bomb under the school, but there was only a 15 percent chance it would go off on any given day?" Tom carped after sitting through a meeting with parents, at Dougherty's request. "Would they leave their kids in school? I thought not." Meanwhile, district-wide budget woes and dwindling enrollment created a more imminent threat to the school's continued existence.

Dougherty had seen to it that he and every principal in the school district had been licensed as a ham radio operator. Each of them had a portable ham radio and knew how to use it. The district was prepared—as prepared as it could be, under the circumstances.

Yes, the school relocation project seemed to be moving along nicely. That was the one bright spot. Tom was glad he had been asked to help.

36

Perpetuity

TOM, 2013

A MIDWEEK EVENING IN LATE JANUARY, approaching 9:30 p.m.: Tom and Kirsten have only just settled back onto their living room couch after clearing the dinner dishes from the coffee table. Tom is in his usual spot at the west end of the sofa with Kirsten's feet in his lap, a blanket covering the two of them. She is fully reclined, enjoying a foot rub from Tom and a movie on TV: a romantic comedy, something she stumbled upon while channel surfing during dinner. Mike is in the oak rocking chair, his head starting to droop.

The TV sits next to the handmade cedar chest tucked in the far corner of the boxcar-shaped living room's western half, the half that had served as the Horning family's dining room when Tom was a boy. Years ago Tom and Kirsten had moved the dining table into the kitchen. They usually ate dinner in front of the TV anyway. The room's eastern half, which Bobbie Horning had set up as the living room, rarely gets used any more except during parties. It is dominated by a white-painted brick fireplace, which itself is dominated by the collection of rocks and minerals and fossils crowding the mantel. It is a glorious collection: unlabeled and unorganized and, as such, so unlike the display Tom had so often admired in Wilkinson Hall at Oregon State. But it is a thing of compelling beauty just the same: snapshots of the Earth in its infancy, its turbulent youth, its dotage. Some specimens Tom has purchased, some he has found, some inherited. A bristle of shiny black epidote spears. A wedge of rhodochrosite like a pink piece of pie set on its side, its polished cut face bursting with crystalline blooms. Orange boils of pyromorphite that Tom had picked up at the Bunker Hill Mine in Idaho. Fossilized horn corals dating from the Devonian period four hundred million years ago that he had found in an eroded limestone hillside in Nevada that first spring after graduating from college. A smooth, tight-grained slab of petrified wood that had belonged to his Grandfather Horning.

Covering the living room walls are Kirsten's own woodblock prints and work by other artists, most of them friends or relatives. Next to the fireplace hangs a pair of hand-carved wooden panels: bas-relief poppies and morning glories worked in eucalyptus wood. The panels, like the cedar chest, are the handiwork of Grandfather Baker. They were Tom's selections back when Bobbie, nearing the end of her life, had asked each of her children what things of hers they wanted.

What he hadn't asked for at that time, what he had hadn't known existed but had found as he and his siblings cleaned out the house following Bobbie's death, was a roughhewn wooden hand. Bobbie had apparently taken up wood carving late in life and had sculpted a life-size left hand modeled on her own. Tom kept it on the top of a tall bookcase against the living room's east wall, in the half of the room where he spent the least amount of time. In life, Bobbie had rarely lain a hand on his shoulder, on his cheek, in his own hand, so that when she did, it felt like a blessing. Once every four or five years, not more than that, he would reach to the top of the bookcase and take down the wooden hand and hold it, savoring the moment. And then he would put it back in its spot. The less often he touched the hand, it seemed, the more power it had.

BOBBIE HORNING HAD BEEN THE THIRD OWNER of the compact, two-story white house at the end of the point, with its warren of small bedrooms and eclectic collection of outbuildings. The Larson family had built it in 1925, carving a buildable lot out of the peninsula's tangle of salal and huckleberry bushes and pine and spruce trees. Tom had long since torn down the old two-car garage, sized for two Model Ts, and shortly after he and Kirsten married had begun work on a new garage. Like the old one, it was two stories tall. But there the similarity ended. At eighteen hundred square feet, it was bigger than his house. He had framed it with four-by-six posts attached to steel moment frames to keep it from flexing, and he spared no expense on the foundation, laying a thick slab of steel-reinforced concrete. "You could launch a Saturn V rocket on it," he liked to boast, though the purpose had less to do with outer space than with the Earth's outer crust. Tom had built the garage to weather a big Cascadia earthquake. Should Kirsten or anyone else be in the garage when it struck, Tom was confident it would stand, allowing her to escape the building and have a good shot at surviving the tsunami. Ten years hence, he still hadn't finished it: hadn't plumbed it, or wired it, or insulated it. He had rigged up some lights and a space heater, powered by an extension cord running out from the house, making the art studio Kirsten had fashioned upstairs useable most of the year.

Downstairs the garage was stuffed with an assortment of tools and toys: table saws, a plastic kayak, two bicycles, a mound of returnable bottles stuffed into plastic bags and waiting to be returned, a shop vac, two kettle barbeques, a riding lawnmower. And a spare refrigerator topped with a cardboard box lined with an electric blanket: a bed for Georgette, the stray cat that had recently adopted Tom and Kirsten.

One end of the garage was walled off into a narrow workroom, though it, too, had become a storage space. Two nylon backpacks, never used and now filmed here and there with mildew, hung from hooks on the wall: Tom and Kirsten's tsunami go-bags. Tom had bought the backpacks almost a decade earlier, filling them with what he considered Boy Scout equipment for an unsupported, two-week campout: water bottles and a backpacker-style water filter, tent, candles and matches, cooking pots and utensils, a first aid kit. Plus a hand-cranked radio, one thing a Boy Scout probably wouldn't take camping. And non-perishable food, mainly beans and rice and jerky. But the rice was gone, as was half the contents of a tube of Neosporin antibiotic ointment, both apparently eaten by Norway rats that had found their way into the garage after a fierce winter storm five years earlier had blown open the big garage doors.

Tom knew that an overhaul of the backpacks, particularly the foodstuffs, was overdue, as were many other chores. What he hadn't neglected was his and Kirsten's evacuation plan, should the tsunami catch them at home. Getting out of the house could be the first challenge, especially if he didn't fix the old pier-and-post foundation, which might settle or slide, jamming doors closed and maybe forcing the occupants to slip out through broken windows. The chimney will collapse; fortunately Tom and Kirsten didn't use that end of the living room much. Once they extricated themselves from the house, they knew exactly the route they would take and how long it would take them; they had practiced it many times. South on Pine Street, then left on Twenty-Fourth Avenue to the highway and across Neawanna Creek. Tom wasn't too worried about getting across the creek; even if the circa-1930 bridge collapsed, as he suspected it would, the creek was no more than chest deep at that point, and the creek bed was rocky enough that they probably wouldn't get bogged down in quicksand-like liquefied sediment. Then north on the highway a few steps before a sharp right on Crown Camp Road. From there it was a scant half mile to the base of the hill. It took them fourteen minutes, Tom at his normal stride, Kirsten walking briskly, to reach the point where the road starts to rise. From Tom's understanding of the latest computer modeling, the first wave is likely to arrive in about twenty minutes, but it will be thirty-five minutes before the flood peaks. That, Tom figured, would give them enough time to grab their bags, start walking, and beat the water to the base of the hills. Once they were there, it shouldn't be hard to stay above the water, whatever its depth. He had seen footage of evacuees in cities on Japan's Sendai plain walking up a hill and, seeing the water continuing to rise, going a little higher. That is the scenario he pictures at the base of the hills east of Seaside.

Whether Mike, Kirsten's father, could escape a tsunami was a different matter. Tom had contemplated several evacuation scenarios with Mike, none of them ending well. Tom and Kirsten couldn't afford to think too hard about Mike and the tsunami, nor about their many elderly and disabled neighbors. Other than to hope that the tsunami didn't come in Mike's lifetime.

Tom's confidence in their ability to escape the tsunami was one of several factors that kept them living in the house by the mouth of the bay. Denial played a role, and nostalgia, and a streak of fatalism. Most of all, beauty: the allure of the estuary. Early in their marriage, Tom had asked Kirsten if she might prefer to move to a new house on the hill above Seaside, safely out of the inundation zone. She had said no. By then she was as beguiled by the setting as he. They might well expire before the tsunami hits, even if they were to live to ripe old ages. Impossible to say which would last longer: the house, or the Hornings.

The question about who would inherit the house was something Kirsten and Tom had talked about from time to time in their thirteen years of marriage. They had no children, though they had plenty of nephews and nieces; should they leave the house to the lot of them? That frankly sounded to Tom like a nightmare, especially after the trouble he and his siblings had themselves had sorting out the house's ownership and occupancy after his mother died. It didn't seem fair to pick just one or two of their siblings' kids to get the house; there was no obvious recipient, no one of them who was more attached to the place than the others. And Tom and Kirsten both hated the thought of an heir turning around and selling it to a developer who would replace the house and cottage with condos or apartments, although that could be prevented with a deed restriction.

But all that discussion was mere noise masking what they both felt was a bigger problem. The house will be destroyed in the next big tsunami: that much they understood. It might happen in their lifetime, and that was a risk both had chosen to accept. They knew they would lose precious mementos, family heirlooms, Kirsten's artwork, but they didn't expect to lose their lives. They were mentally prepared to huddle in the woods for days or even weeks, to bounce between friends' and family members' houses for several months until they reestablished a home for themselves in whatever was left of Seaside or Astoria.

But what about the next occupants? Would they have the wherewithal to plan their evacuation as Tom and Kirsten had? Would they have young children who might need to be carried, or would they be old and unable to move quickly? Would they get around to fixing the foundation?

Then Tom proposed an idea to Kirsten: how about leaving the house to the land trust?

Tom had joined the board of the North Coast Land Conservancy shortly after moving back to Seaside. It operated like a small-scale, regional version of The Nature Conservancy, acquiring or otherwise conserving undeveloped land, or land that had already been grazed or farmed or logged but could be rehabilitated, to preserve some of its original wildness and wildlife habitat. The land trust's values overlapped his own. He still mourned the disappearance of the blackberry thickets and wild crabapple trees that had dominated his neighborhood when he first moved in, and their replacement by more and more houses. He hated to see Gearhart's cranberry bogs converted to golf courses. He missed the estuary's once-vigorous runs of native salmon, whose survival now hung by a thread. He missed the song of the western meadowlarks in the uplands east of the beach, the way they would return his whistle. In recent years he had become better attuned to the broader ecological imperatives that drove the land trust's decisions. But he wasn't too proud to admit it: he had joined the movement because he didn't like change, not the kind of change that muscled out meadowlarks. And he didn't like ugly.

If they were to leave their house to the land trust, Tom said to Kirsten, spelling out his idea, the trust could sell it to the city and use the money from the sale to conserve more land elsewhere. Meanwhile, the city could tear down the house and other buildings and create a neighborhood park. It would mirror Neawanna Point, the undeveloped stretch of salt marsh across the creek, already preserved by the land trust; it was the view out the Hornings' kitchen window. The city could even fix up the dock, giving the public a place to fish and crab. Tom and Kirsten could die knowing that their estate would be used to help preserve the kinds of places they valued most, in perpetuity, and knowing that no one would die in their house when the tsunami arrived.

Drawing up wills was another item on the to-do list, probably a little overdue, with Tom in his late fifties and Kirsten just a few years behind him. With no children, no particular health problems, and both of them employed, there had seemed no urgency. The plan was to first discuss the concept with members of the board's fund-raising committee, giving them some needed practice in soliciting planned gifts of this kind.

That discussion had never happened, or hadn't yet. At 5:30 the morning of December 5, a logger and land trust supporter on his way to work on Highway 101 glanced, as was his habit, across the river to the land trust's 360-acre Circle Creek preserve and the two-story house that served as its headquarters. But this particular morning he saw, in the darkness, an odd glow beyond the leafless alders. He stopped, looked harder, then—he had no cell phone—headed to a nearby friend's house and pounded on the door, shouting his own name and "911." The Seaside Fire Department responded immediately, backed by firefighters and equipment from the Cannon Beach, Gearhart, Warrenton, and Lewis and Clark fire departments. But it made no difference. The house and everything in it had burned to the ground: twenty-five years of maps, photographs, paper files, computers and their data, all gone. Actually not all the computers had burned; some had apparently been stolen before the house was set on fire, according to fire investigators later piecing together the evidence.

Perpetuity: it was something the staff and volunteers of the land trust talked about all the time. It was their promise as land stewards: to conserve the land in perpetuity, forever. Not wood-frame houses; no one expects them to last forever. But no one had anticipated a fire, not now, not here. No one ever does.

What did perpetuity mean anyway? Tom had just finished a report for the land trust documenting the way Cascadia earthquakes and tsunamis periodically reshape the Clatsop Plains dune complex between Gearhart and the Columbia River. Parallel dunes, each one memorializing the last tsunami. A landscape in a continuous, active state of change.

It wasn't really true that Tom didn't like change. He had spent his life witnessing geologic change and its aftermath: erupting volcanoes and seeping magma, sand deposition from sweeping tsunamis, tracks of landslides, and the circumstantial evidence of seafloor transformed into mountaintops. People use the term "geologic time" to mean slow, too slow for humans to quite comprehend, but geology isn't always slow. Every moment, tectonic plates creep forward incrementally, and the continent's edge rises. Every few hundred years, in the town where he had lived most of his life, the land is re-leveled and the sea takes new liberties.

A few years earlier, scientists from a project called the Plate Boundary Observatory, funded by the National Science Foundation, had discussed with the land trust establishing a monitoring station on the trust's Circle Creek property. The project uses a variety of high tech instrumentation to precisely measure on-going deformation of the Earth resulting from the constant motion of tectonic plates, particularly the Pacific and Juan de Fuca and North American plates in the western United States. The project's staff measures change, but they liked the sound of *perpetuity*—of installing their very expensive instrumentation at a site unlikely to ever be sold and have a house built on it. Tom had been on hand as the observatory's crew drilled a hole seven hundred feet deep in the land trust office's front yard, through the dirt and sediments and cobble, through layers of sandstone and Grande Ronde basalt and Frenchman Springs basalt—rock formed some fifteen million years ago from lava flowing out of volcanic hotspots hundreds of miles to the east. At the bottom of the hole, they cemented in place a strain gauge, one of a network of such gauges that monitor in what direction the ground is compressing and extending. Tom had gone online, to the project's website, and had seen the results: a distinct movement northeastward, as the North American Plate pushes to the west-southwest and the Juan de Fuca Plate drifts east and north, toward Alaska.

Change on Tom's own property was slow and fast as well. Slow: the rise of the land itself, too slow to observe in even one human lifetime, slower than the growth of the spruces and pines on the shore, faster seemingly than the pace of his remodeling projects. One day his land would change very fast. It would probably subside a foot or even three feet during the earthquake, and then the tsunami would sweep in and wipe it clean not only of buildings but of most of the vegetation. The century-old Sitka spruces clustered behind the garage might withstand the tsunami but would soon die from the saltwater inundation of their roots. In the years that would follow, winter storm waves would wash the land, smoothing it and turning it into a mirror of Neawanna Point: a tidal wetlands, swathed in saltgrass and pickleweed, inundated and drained twice daily by the tides. The dying spruce would rot and topple and become bedding for new plants colonizing the salt marsh. And the cycle would continue, as it had for millennia, while a new town—its citizens chastened by Mother Nature and a new understanding of statistical probability—was rebuilt at Seaside, probably on ground a little higher, a little farther to the east.

That question Tom had asked Kirsten, if she wanted to sell the house and move to a new one up the hill: perhaps it had been merely rhetorical. Could he really leave this place? More than his home, it was his milieu. The house itself had survived more storms than he, eighty-seven years' worth. Both he and the house were aging, and neither would last forever. It might not be a tsunami that destroyed the house; the fire at the land trust office had been a stark reminder of the many faces of calamity, and of its utter unpredictability. What made a tsunami seem both realer and less real than other types of disasters: on any given day, a tsunami was extremely unlikely to occur, the chances infinitesimal. But that it *would* happen someday was absolutely certain, as certain as human mortality.

The fire at the land trust office was, in a way, the best-case scenario for a devastating disaster: every*thing* is lost, but every*body* survives. But the nexus of that disaster was a single building; the tsunami's target was a 620-mile-long coastline, and more: Japan's Pacific coast and other shorelines ringing the North Pacific. Tom understood that *no deaths* from the tsunami was an unrealistic goal. But *fewer* deaths was entirely within reach. *If only we have the will.*

It's nearly midnight: Tom is asleep, his long legs still propped on the coffee table, his head propped by a pillow atop the sofa cushion, his glasses still propped on his nose. Kirsten is asleep too, but now she rouses, carefully disentangling her feet from Tom's lap. She has learned to not wake Tom, even in the service of getting him off the couch and into bed; he tends to growl irritably—a reflexive reaction to being pulled from the sweet delirium of the first hours of sleep. It's best, she's learned, to just leave him be. The oak rocker is empty; Mike has already retreated to his bedroom, formerly Tom's home office, formerly Bobbie Horning's bedroom. Kirsten gets up, tucks her end of the blanket back around Tom, turns off the TV and the lights, and climbs the stairs to their bedroom.

Outside the neighborhood is dark; clouds obscure any light that might otherwise shine from the moon or the stars. The Sitka spruce boughs rub the side of the garage, creating a shushing sound. Georgette slips into the garage through the cracked-open side door: back, perhaps, from a late prowl for mice among the dead stalks and desiccated vines of the overgrown garden. The light in the upstairs bedroom window clicks off. In the estuary, the tide is out, and the mudflats yawn black and wet. But here and there, house lights and headlights glimmer in the dark.

Epilogue

IN JUNE 2013, JUST THREE WEEKS after a magnitude 8.3 earthquake rocked the Pacific Plate under the Sea of Okhotsk north of Japan, Cannon Beach Elementary School closed for good. Superintendent Doug Dougherty had hoped to keep it open until the new tsunami-safe campus he envisioned in Seaside could be built. But declining enrollment and tight budgets forced his hand. Plans were made to bus kids from Cannon Beach to Seaside Heights Elementary, six miles to the north and seventy feet higher on the hillside, beginning the following September. Some parents were deeply disappointed at losing their small town's school, and they redoubled their efforts to find a site in or near Cannon Beach where they could build a new charter school. Meanwhile the City of Cannon Beach began conversations with the school district about acquiring the abandoned elementary school's scenic shoreline campus, which they planned to turn into a city park after demolishing the school buildings.

The same week that tearful goodbyes were playing out at Cannon Beach's old school—just days before a magnitude 6.5 quake off the Pacific coast of Nicaragua and a magnitude 5.8 temblor ninety miles south of Mexico City—Oregon's Department of Geology and Mineral Industries provided the newspapers in Astoria and Seaside with a preview of its new tsunami inundation maps for Clatsop County. Even in communities well aware of their vulnerability to a tsunami, the new maps were a bit of a shock. In Seaside, the orange zone—the worst-case scenario for inundation from a *distant* tsunami—included the entire business district, every school but Seaside Heights, and all but a few homes clustered on the hillside east of town. In a very large *local* tsunami, according to the new maps, even Seaside Heights and the hospital could end up under water. The map for the community of Warrenton, arrayed on the flat coastal plain south of the mouth of the Columbia River, was overwhelmingly yellow and orange, with just a few strips of green, indicating high ground, on the highest dune ridges. The picture was even bleaker for the town of Gearhart. Unique among DOGAMI's Oregon coast tsunami inundation maps, the Gearhart map had no patches of solid green but, rather, green cross-hatching in a few spots to indicate "optional high ground assembly areas"—dune ridges that, geologists figured, would stay dry in all but the very biggest tsunami. But in an XXL tsunami—the kind generated by, say, a magnitude 9.1 quake—even those spots could be inundated.

"We're toast," was the response of one Gearhart city councilor during a presentation by DOGAMI geologist Rachel Lyles Smith. "We're toast."

"I can't sugarcoat it," Smith responded. In a worst-case tsunami, she said, "there will be a lot of fatalities in this area."

But receipt of the new maps didn't translate into support for the Seaside School District's $128.8 million bond measure to fund construction of a new consolidated school campus above the tsunami inundation zone. The measure failed decisively, 61 percent to 38 percent, in the November 2013 election. Supporters of the bond measure had focused their campaign on children's safety and on the advanced technology that would be built into state-of-the-art classrooms; the slogan on lawn signs read "SAFE: Save A Future for Education." But the bottom line for many *no* voters seemed to be the cost. Even some supporters were having a hard time swallowing the increase in property taxes that construction of the new campus would have required. Which sent Dougherty and the school board back to the drawing board to consider other options, such as building a smaller campus on the hill for starters, one to house just the elementary school students.

In Oregon, all elections are conducted by mail. Between late October, when voters received their ballots, and Election Day, when ballots were due and would be counted, no fewer than four earthquakes large enough to be considered "significant" by the U.S. Geological Survey occurred in the Pacific Basin, off Mexico, Japan, Taiwan, and Chile, ranging from 6.3 to 7.1 in magnitude. None of them made the local papers.

Tom Horning was disappointed in the school bond measure failure—he had been a member of the SAFE campaign committee—but he took the loss in stride. One battle lost, but the war—as he saw it—for a safe community would rage on. He began dreaming about buying a piece of land in the forest and building a survival cabin: maybe six hundred square feet, loaded with food and propane and other essentials, a place where he and Kirsten could hole up should the need arise. *Maybe when people see what the geologist is doing,* he mused, *they might change their views on the hazard.*

Meanwhile he began making arrangements to get the two-story garage behind his house wired and plumbed and Kirsten's upstairs studio finally insulated and the walls finished. The builder expected to start the work before the first of the year.

Selected References

BOOKS AND ARTICLES IN PRINT

Adams, J. "Paleoseismicity of the Cascadia Subduction Zone—Evidence from Turbidites off the Oregon-Washington Margin." *Tectonics*, vol. 9, no. 4 (August 1990): 569–83.

Ando, M., and E. I. Balazs. "Geodetic Evidence for Aseismic Subduction of the Juan de Fuca Plate." *Journal of Geophysical Research*, vol. 84 (1979): 3023–28.

Atwater, Brian F. "Evidence for Great Holocene Earthquakes Along the Outer Coast of Washington State." *Science*, n.s. vol. 236 (May 22, 1987): 942–44.

Atwater, Brian F., and David K. Yamaguchi. "Sudden, Probably Coseismic Submergence of Holocene Trees and Grass in Coastal Washington State." *Geology*, vol. 19 (July 1991): 706–9.

Atwater, Brian F., et al. *The Orphan Tsunami of 1700*. Seattle: University of Washington Press, 2005.

Atwater, Brian F., et al. "Summary of Coastal Geologic Evidence for Past Great Earthquakes at the Cascadia Subduction Zone." *Earthquake Spectra*, vol. 11, no. 1 (1995): 1-18.

Atwater, Tanya. "Implications of Plate Tectonics for the Cenozoic Tectonic Evolution of Western North America." *Geological Society of America Bulletin*, vol. 81 (December 1970): 3513–36.

Blackett, P. M. S., Sir Edward Bullard, and S. K. Runcorn, eds. *A Symposium on Continental Drift*. London: The Royal Society, 1965.

Bommelyn, Loren. "Test-ch'as (The Tidal Wave)." In *Surviving Through the Days: Translations of Native California Stories and Songs*, edited by Herbert W. Luthin, 67–76. Berkeley: University of California Press, 2002.

Bryant, Edward. *Tsunami: The Underrated Hazard*. Cambridge: Cambridge University Press, 2001.

Clague, John J., et al. (compilers). *Penrose Conference 2000: Great Cascadia Earthquake Tricentennial*. Oregon Department of Geology and Mineral Industries Special Paper 33 (2000).

Clarke, Samuel H., and Gary A. Carver. "Late Holocene Tectonics and Paleoseismicity, Southern Cascadia Subduction Zone." *Science*, n.s. vol. 255, no. 5041 (January. 10, 1992): 188–92.

Committee on the Alaska Earthquake of the Division of Earth Sciences, National Research Council. *The Great Alaska Earthquake of 1964: Human Ecology.* Washington, D.C.: National Academy of Sciences, 1970.

Committee on the Alaska Earthquake of the Division of Earth Sciences, National Research Council. *The Great Alaska Earthquake of 1964: Oceanography and Coastal Engineering.* Washington, D.C.: National Academy of Sciences, 1972.

Connolly, Thomas J., et al. *Human Responses to Change in Coastal Geomorphology and Fauna on the Southern Northwest Coast: Archaeological Investigations at Seaside, Oregon.* Eugene, OR: Department of Anthropology and Oregon State Museum of Anthropology, University of Oregon (April 1992).

Darienzo, Mark E., and Curt D. Peterson. "Episodic Tectonic Subsidence of Late Holocene Salt Marshes, Northern Oregon Central Cascadia margin. *Tectonics,* vol. 9, no. 1 (February 1990): 1–22.

Dickinson, William R. "Global Tectonics." *Science,* vol. 168 (June 5, 1970): 1250–59.

Dickinson, William R. "The New Global Tectonics." *Geotimes,* vol. 15, no. 4 (April 1970); 18–22.

Emmons, William H., et al. *Geology: Principles and Processes,* 5th ed. New York: McGraw-Hill Book Company Inc., 1960.

Glen, William. *The Road to Jaramillo: Critical Years of the Revolution in Earth Sciences.* Stanford, CA: Stanford University Press, 1982.

Gonzales, Laurence. *Deep Survival: Who Lives, Who Dies, and Why.* New York: W.W. Norton, 2003.

Griggs, Gary B. "The First Ocean Floor Evidence of Great Cascadia Earthquakes." *Eos,* vol. 29, no. 39, (September 27, 2011): 325–36.

Horning, Tom. "Summary of Eyewitness Observations from 1964 Alaska Tsunami in Seaside, Oregon." In *Seaside, Oregon Tsunami Pilot Study— Modernization of FEMA Flood Hazard Maps.* Seattle: NOAA OAR Special Report, NOAA/OAR/PMEL, 2006.

Kerr, Richard A., "Faraway Tsunami Hints at a Really Big Northwest Quake." *Science,* vol. 267 (February 17, 1995): 962.

Kroeber, A. L. *Yurok Myths.* Berkeley: University of California Press, 1976.

Kroeber, Theodora. *The Inland Whale.* Bloomington: Indiana University Press, 1959.

Ludwin, Ruth S., et al. "Dating the 1700 Cascadia Earthquake: Great Coastal Earthquakes in Native Stories." *Seismological Research Letters,* vol. 76, no. 2 (March/April 2005): 140–48.

Maxwell, Arthur E., et al. "Deep Sea Drilling in the South Atlantic." *Science,* vol. 168, no. 3935 (May 29, 1970): 1047–59.

McCoy, Roger. *Ending in Ice: The Revolutionary Idea and Tragic Expedition of Alfred Wegener.* New York: Oxford University Press, 2006.

Oreskes, Naomi, ed. *Plate Tectonics: An Insider's History of the Modern Theory of the Earth.* Boulder, CO: Westview Press, 2003.

Peterson, Curt D., et al. "Recurrence Intervals of Major Paleotsunamis as Calibrated by Historic Tsunami Deposits in Three Localities: Port Alberni, Cannon Beach, and Crescent City, along the Cascadia Margin, Canada and USA." *Natural Hazards,* vol. 68, issue 2 (September 2013): 321–36.

Plafker, George. "Tectonic Deformation Associated with the 1964 Alaska Earthquake." *Science,* n.s. vol. 148, no. 3678 (June 25, 1965): 1675–87.

Pope, Daniel. *Nuclear Implosions: The Rise and Fall of the Washington Public Power Supply System.* New York: Cambridge University Press, 2008.

Priest, George R., et al. "Tsunami Hazard Assessment in Oregon." International Tsunami Symposium, 2001 *Proceedings,* Paper R-3: 55–65.

Raff, A. D., and R. G. Mason. "Magnetic Survey off the West Coast of North America, 40° N Latitude to 52° N Latitude." *Bulletin of the Geological Society of America,* vol. 72 (1961): 1267–70.

Rubinger, Richard. *Popular Literacy in Early Modern Japan.* Honolulu: University of Hawai'i Press, 2007.

Satake, Kenji, et al. "Time and Size of a Giant Earthquake in Cascadia Inferred from Japanese Tsunami Records of January 1700." *Nature,* vol. 379 (January 18, 1996): 246–49.

Schwarzbach, Martin. *Alfred Wegener, the Father of Continental Drift.* Madison, WI: Science Tech, 1986.

Strahler, Arthur N. *The Earth Sciences.* New York: Harper & Row, 1963.

Tsuji, Y. "Konohi no enchi tsunami kiroku WANTED [Shimei tehai shimau] (Wanted: A Record of Tsunamis from Distant Places on a Particular Day [My Search for Information])." *Rekishi Jishin (Historical Earthquakes),* mtg. 7 (1990), 47–59.

Tsuji, Y. "Shoutsunami no shiryou (Historical Materials about Small Tsunamis)." *Rekishi Jishin (Historical Earthquakes)*, vol. 3 (1987), 220–38.

Vine, F. J., and D. H. Matthews. "Magnetic Anomalies over Oceanic Ridges." *Nature*, vol. 199, issue 4897 (September 7, 1963): 947–49.

Waterman, T. T. *Yurok Geography*, facsimile of 1920 edition. Trinidad, CA: Trinidad Museum Society, 1993.

Wegener, Alfred, tr. John Biram. *The Origin of Continents and Oceans*, 4th ed. (1929). New York: Dover Publications, 1966.

Williams, C. A. *Madingley Rise and Early Geophysics at Cambridge.* London: Third Millennium Publishing Ltd., 2009.

Wilson, J. Tuzo. "Transform Faults, Oceanic Ridges, and Magnetic Anomalies Southwest of Vancouver Island." *Science*, n.s. vol. 150, no. 3695 (October 22, 1965): 482–85.

Yamaguchi, David K. "New Tree-Ring Dates for Recent Eruptions of Mount St. Helens." *Quaternary Research*, vol. 20 (1983): 246–50.

Yamaguchi, David K., and Richard P. Hoblitt. "Tree-ring Dating of pre-1980 Volcanic Flowage Deposits at Mount St. Helens, Washington." *GSA Bulletin*, vol. 107, no. 9 (September 1995): 1007–93.

Yamaguchi, David K., et al. "Tree-ring Dating the 1700 Cascadia Earthquake." *Nature*, vol. 389 (Oct. 30, 1997): 922–23.

Yeats, Robert S. *Living with Earthquakes in the Pacific Northwest.* 2nd ed. Corvallis: Oregon State University Press, 2004.

Yeats, Robert S. "Summary of Symposium on Oregon's Earthquake Potential Held February 28, 1987, at Western Oregon State College in Monmouth." *Oregon Geology*, vol. 49, no. 8 (August 1987): 97–98.

Online Sources

Goldfinger, Chris. "Short Memories" and "Why Was It a Surprise?" Atquake (blog). http://atquake.wordpress.com/

Goldfinger, Chris, et al. "Turbidite Event History—Methods and Implications for Holocene Paleoseismicity of the Cascadia Subduction Zone." U.S. Geological Survey Professional Paper 1661–F (2012). http://pubs.usgs.gov/pp/pp1661f/.

Oregon Department of Geology and Mineral Industries. www.oregongeology.org.

Plate Boundary Observatory. http://pbo.unavco.org/

Wood, Nathan J., Jamie Ratliff, and Jeff Peters. "Community Exposure to Tsunami Hazards in California." U.S. Geological Survey Scientific Investigations Report 2012–5222 (2012). http://pubs.usgs.gov/sir/2012/5222/.

Wood, Nathan, and Christopher Soulard. "Variations in Community Exposure and Sensitivity to Tsunami Hazards on the Open Ocean and Strait of Juan de Fuca Coasts of Washington." U.S. Geological Survey Scientific Investigations Report 2008–5004 (2008). http://pubs.usgs.gov/sir/2008/5004/.

Wood, Nathan. "Variations in City Exposure and Sensitivity to Tsunami Hazards in Oregon." U.S. Geological Survey Scientific Investigations Report 2007 5283 (2007). http://pubs.usgs.gov/sir/2007/5283/.

Unpublished Sources

Carver, Deborah H. "Native Stories of Earthquakes and Tsunamis Prepared for Redwood National Park, Orick, California, 1998."

Crandell, Joan. "The Indian Island Massacre: An Investigation of the Events that Precipitated the Wiyot Murders." Master's thesis, Humboldt State University, 2005.

Darienzo, Mark. "Late Holocene Geologic History of a Netarts Bay Salt Marsh, Northwest Oregon Coast, and its Relationship to Relative Sea Level Changes." Master's thesis, University of Oregon, Eugene, 1987.

Dickinson, William R. "Report on the Second Penrose Conference: The Meaning of the New Global Tectonics for Magmatism, Sedimentation, and Metamorphism in Orogenic Belts," December 15-20, 1969."

Fiedorowicz, Brooke K. "Geologic Evidence of Historic and Prehistoric Tsunami Inundation at Seaside, Oregon." Master's thesis, Portland State University, 1997.

Goddard, Pliny E. "Tolowa Tales and Texts 1902-1911" (manuscript). Berkeley: California Indian Library Collections (distributor), 1992.

Horning, Tom. "The Geology, Igneous Petrology, and Mineral Deposits of the Ataspaca Mining District, Department of Tacna, Peru." Master's thesis, Oregon State University, 1988.

Seaside Tsunami Advisory Group. "Tsunami Strategic Investment Plan, Seaside, Oregon." February 2009; updated May 2012.

See, Paul. "Wisdom of the Elders" (unfinished manuscript).

Acknowledgements

My thanks go to the following scientists and others who gave generously of their time for sometimes extended interviews and manuscript review: Brian Atwater, Tanya Atwater, Alfred Aya, Antonio Baptista, Doug Barker, Loren Bommelyn, Deborah Carver, Gary Carver, Kevin Cupples, Mark Darienzo, William Dickinson, Doug Dougherty, Brooke Fiedorowicz, Chris Goldfinger, Kirsten Horning, Neal Maine, Alan Nelson, Steve Olson, Curt Peterson, George Plafker, George Priest, Kenji Satake, the late Paul See, Kunihiko Shimazaki, Al Smiles, Tom Stephens, Kazue Ueda, Fred Vine, Mark Winstanley, Rob Witter, David Yamaguchi, and Bob Yeats. My sincere apologies to the many others whose work contributed to the Cascadia puzzle, who deserve chapters and whole books of their own but who I didn't feature here out of respect for the reader's limits. A few of their papers central to the scope of this book are listed in "Selected References."

Many thanks to friends and family for early reading of first chapters, thoughtful review of the manuscript, a quiet place to write, and other essential support: Betsy Ayres, Kathleen Fitzgerald, Darcy Henderson and Mike Urness, Donna Henderson and Rich Sutliff, Randall and Jeanne Henderson, Barbara Lee and John Costello, David Markewitz, Jim and Suzanne Moody, and Donna Scurlock and Mike Weinstein. Particular thanks to Randall, who got the ball rolling and was a cheerleader throughout, and to Donna Henderson for her wordsmithing and encouragement.

I am indebted to Yoko O'Brien for translation assistance in Oregon and to Satoko Musumi-Rokkaku for translation and generous hospitality in Yokohama.

Deep thanks to Playa for a writing residency in fall 2011 that was essential to my completion of the book. Assignments from *Oregon Quarterly* ("What Happens When It Happens Here?" and "Big Wave, Small World") helped support research for the book.

I am grateful for editor Elizabeth Lyon's advice and encouragement, and to Ursula Le Guin for fact checking and cheering an early chapter draft. My thanks to author Rob Guth and publicist Robin Carol for their efforts on the book's behalf.

Gary Durheim and Les Neitzel reviewed portions of the text involving logging practices and chainsaw use and improved the manuscript with their wordsmithing and fact checking; thanks, guys.

The staff of OSU Press is a pleasure to work with; thank you.

The maps are the work of Erin Greb Cartography; thank you, Erin.

I join surfers, geologists, and others in offering my appreciation, written and unwritten, to proprietors of The Stand in Seaside.

The World Wide Web is an invaluable research tool, but there is still no substitute for personal research assistance of the kind I received from staff at the University of Oregon Science Library, the Eugene Public Library, the Indian Action Council, the Tokyo Foundation, and the *Willapa Harbor Herald*.

I am grateful to Charlie Zennaché for many things, including love, unflagging encouragement, and literary patronage, and to our kids who have encouraged and—as they pursue their own passions—inspired me.

My greatest thanks go to Tom Horning for his candor and the hours and days he spent talking with me and reviewing drafts. Tom, you have been a patient teacher and became a treasured friend; your willingness to share your story allowed me to undertake what became a very rewarding project. May it make a difference.

Index

Adams, John, 119, 222–24
Aleutian Megathrust fault, 39
Alsea Bay, 93–94
Ando, Masataka, 100–102
anti-war activism, early 1970s: in Japan, 172; in the United States, 94, 110
Ataspaca Prospect (Peru), 134–36
Atwater, Brian: background, 98; collaboration with David Yamaguchi, 86–90, 121–29; marsh stratigraphy research, 95–104; Penrose Seaside conference, 233–34; standard penetrometer studies, Columbia River, 157
Atwater, Tanya, 68–69, 72–76, 92, 110, 183
Aya, Alfred, 138–47, 182

Baker, Frederick Storrs, 19, 42–44, 89
Balazs, Emery, 100–102
Baptista, Antonio, 159–62, 164–65, 184–85
bark beetle, 133
barnacle, 16–17, 37–38
Big Lagoon, 106, 151–52, 192
Blackett, P.M.S., 53–54, 60
Bommelyn, Loren, 206–8
Buckley, Thomas, 205
building code, Oregon, 235, 244
Bullard, Edward, 55, 58

California Division of Mines and Geology, 107
California Indian Conference (1994), 207
Cannon Beach Elementary School, 244; closure, 305; effort to relocate, 247–50, 252, 295; tsunami drills, 245–47, 251; vulnerability to earthquake and tsunami, 244–47.
Cannon Beach, Oregon: COwS, 142–43; impact of 1964 tsunami, 140, 179, 244–45; paleotsunami research in, 247–49, 251–52; tsunami evacuation mapping, 241; tsunami preparedness, 139–47; vulnerability to tsunami, 2
Carver, Deborah, 191–208
Carver, Gary: background, 108–10; research on Cascadia fault traces, 105–8, 110–112; Monmouth conference, 118–20; collaborations with Deborah Carver, 191–96, 210–11
Cascades Volcano Observatory, 124, 189
Cascadia Margin Meeting, Salishan (1979), 92–93, 184
Cascadia Megathrust earthquake (1700): dendrochronology research on, 122–33, 186–90; radiocarbon dating of, 167–68, 210–11; use of Japanese historical records to identify, 166–69, 171–77
Cascadia Subduction Zone, 1, 75, 92, 124, 184; early speculation about seismicity of, 91–95, 98–101, 111, 136; magnitude and frequency interval estimates of past earthquakes, 210, 284–86; magnitude and timing projections for future earthquake, 182–83, 264,

315

Author photo by Jack Liu

Bonnie Henderson is the author of *Strand: An Odyssey of Pacific Ocean Debris* (OSU Press)—an Oregon Book Award finalist and one of the *Seattle Times*' Best Books of 2008, as well as two hiking guidebooks. She has been a newspaper reporter and editor, an editor at *Sunset* magazine, and a writer for a number of magazines including *Backpacker, Ski,* and *Coastal Living.* Currently a freelance writer and editor focusing on the natural world, Henderson divides her time between the Oregon coast and her home in Eugene, Oregon.